Foreword by Bill McKibben

THE GLOBAL AWAKENING SERIES

How We're Fixing What We've Almost Broken, And How You Can Help

Volume 3

GLIMPSES OF ONENESS

Facets of the Unity Perspective

Lee Temple

SHINING GOLDEN SUNS, LLC
CRESTONE, COLORADO

GLIMPSES OF ONENESS

First Edition, 2015

© 2015 Lee Temple, Shining Golden Suns, LLC

All Text © Lee Temple, Shining Golden Suns, LLC except as noted

Copyright © of all photographs/illustrations belongs to the photographer/illustrator except as noted otherwise.

ISBN 978-1-941306-06-2

eISBN 978-1-941306-05-5

Produced by Lee Temple and Shining Golden Suns, LLC

Designed by Lee Temple, Paul Cash, Anne Kilgore, Molly Rowan Leach, and Elise Rudoff.

Published by Shining Golden Suns, LLC

For further information about The Global Awakening Series, Lee Temple, and/or Shining Golden Suns, LLC, please visit our website at www.primamundi.com

The electrical energy used by all collaborators to build this document in various locations has been entirely offset by 100% Genuine Carbon-Neutral San Luis Valley Solar Power generated on-site at the author's Wingspread Sustainable Homestead, Crestone, Colorado.

According to a globally representative 2010 Pew Research Poll, 84% of people polled believe there is a global climate crisis, compared with an annually declining percentage of 70% of Americans. The study also found that majorities in most countries believe that human activity is responsible.[1] More recently, after 2012's super-hurricane Sandy, and unprecedented heat that caused American farmers to lose half their corn and 59% of their pastures to the searing drought, an Associated Press poll found that almost 80% felt that Washington should act decisively against climate change, indicating a radical shift in American public opinion in just two years.[2]

Although we've known about it for decades, global greenhouse gas levels continue their unchecked exponential growth. Global carbon emissions continued to escalate during the recent financial downturn, and now their growth is accelerating again. The current **global carbon footprint** approached 36 billon metric tons per year in 2013 (up 2.1% from 2012). We've also passed the ignominious milestone of 400.0 parts per million (ppm) in the atmosphere, far above the 350 ppm consensus target, and over 60% above the 1997 Kyoto Protocol's baseline levels. The International Panel on Climate Change (IPCC) **says we've entered the tipping point,[3]** the time after which no human activity will be able to forestall catastrophic global climate change, although human carbon emissions likely won't peak until around 2030 at the earliest.

Governments remain unable to muster the political will to address this greatest crisis of any human era effectively. As their gridlock wastes more precious time, we have more than sixty million refugees worldwide, many of whom have been displaced by climate change—more than twenty million affected by the July 2010 Pakistani monsoon alone. Some estimates place the climatically disenfranchised at two hundred million by 2050.[4]

Clearly, our climate crisis playbook/toolkit is incomplete and disintegrated. **What's missing is a larger sense of selfhood**—a new ethics/worldview founded in a sensibility of wholeness and global responsibility, one that values the sources and sustenance of our existence as much as ourselves and our personal wants and needs. The *Global Awakening* series provides such a comprehensive worldview.

THE *GLOBAL AWAKENING* SERIES

The *Global Awakening* project integrates widely accepted scientific, spiritual, and environmental perspectives to reveal our profound collective journey from the Big Bang to the present day—so we can understand, at the deepest core level, who we truly are, what we're made of, and how we've come face to face with the greatest human challenge of all time—global climate change. In so doing, it facilitates a large-scale human awareness shift into a new, direct experience of the universe's intrinsic unity. This enlightening transformation—as essential to successfully addressing climate crisis as the advancement of green technology or governmental initiatives—inspires broad-based, compassionate, win-win behavior that promotes lasting healing for Earth and her children. For more information, please visit www.primamundi.com.

For Carol, Whose Love and Inspiration
Made this Book Possible

Dedication

This work is dedicated to all the great beings on whose shoulders I stand;
those in my biological, educational, Earth-healing and spiritual families,
and especially to all those beings, animate and inanimate,
sentient all the same, who have played and will continue
to play an important role in the unfoldment
of my continuing evolution in all these areas.

I also dedicate it to all those great beings whose lives will be affected by it.
May this book inspire you to go on to do great things in your own way!

And I dedicate it to this miraculous little blue planet Earth,
our great Mother, who nurtures us all with her great, great Love.
May we all live to see the day that Her fever is healed and true harmony
emerges among all Her sentient and insentient-being families.
In Unity We Stand!

The Global Awakening Series

C O N T E N T S

Volumes in the Series:

Volume 3 Includes:

FOREWORD BY BILL MCKIBBEN

We are in a tremendous fix. We burn a lot of coal and gas and oil on a daily basis in this world—as a result, the atmospheric concentration of CO_2 has zoomed past the maximum safe level of 350 parts per million, and now nudges 400 ppm. As a result, the Arctic, and most of the other ice on the planet, is melting swiftly; as a result, the oceans are 30% more acidic than they were a generation ago; as a result, we see a steady and devastating increase in both flood and drought.

This is the first truly global problem that we've ever faced. The sources of the trouble aren't found everywhere—many people on Earth burn little or no fossil fuel—but the effects can now be found in every corner of the planet. In fact, many of the hardest hit places are ones that have done little to cause the crisis.

Lee with Bill McKibben and Tim DeChristopher (seated from right to left) at Telluride's MountainFilm Festival, May 2011, shortly before Tim started serving time for his 2008 Utah oil/gas lease auction protest, documented in the film "Bidder70."

We don't lack for answers: Renewable energy is now a very real possibility, not just a hope. But the forces of the status quo—especially the big energy companies—do everything they can to keep us from getting there. So the world really does need to unify, in order to stand up to their millions. We'll never outspend Exxon, but unified we can find other currencies to work in: passion, spirit, creativity. It's been so gratifying to watch millions come together for demonstrations around the world—but this movement needs to get much bigger, to include as much of humanity as it possibly can.

Big crises can lead to big, powerful changes. Let's hope we can make that happen here!

AUTHOR'S PREFACE

As time goes by, we see steadily increasing devastation from global climate change all around the world. We also see clearer and clearer evidence of our role in its acceleration, and ever more convincing proof that our negative effect on climate is accelerating as well. We can't escape the troubling conclusion that, if we want to deliver a habitable planet to our grandchildren, we must unite quickly in actions across the globe to make a decisive difference—by changing our destructive behaviors now, while we still can.

Uniting in *action* to heal our world is urgent now, and will become even more so in the next few years. I am a hundred percent certain that we can and will achieve it.

Contributing positively to this global healing transformation is so important to me that I've devoted the past eight years to conceiving and developing what I call the "Global Awakening" series—a massive work that addresses all the aspects of this crucial topic. I hope it will help us effectively engage this task without delay—to make changes that must be in place soon to elude a crucial tipping point that otherwise will be upon us.

The *Global Awakening* series is based on my own personal experience of awakening into the perspective of oneness, and finding ways to apply that perspective successfully in how I live and work. There are indestructible unifying bonds between the various parts that compose each and every one of us, between all of us as humans, and between our species and our world. Contrary to divisive propaganda, these bonds are right here for each one of us to experience. The ways we can employ them to heal our world are also right here, ready for each of us to use.

Our tremendous recent progress through technological advancements makes it easier than ever for us to forget or ignore three important shared truths of our essential interdependence:

- We are all products of our world.
- Life on our precious Mother Earth is in mortal danger.
- We still have the capacity to join together and make things better.

This overview volume—*Glimpses of Oneness*— presents a series of brief essays that sketch various facets of my own life's awakening. These often playful fables share inspiring nuances of the unity perspective, as seen in a daily life context that is easily recognizable to all. Each represents a pivotal moment in or aspect of my evolutionary development.

Some are experience-based, and are presented more or less chronologically as they appeared during a fifty-year arc of my life's trajectory. Others chronicle some of the forms that have emerged for me as a direct result of these important oneness experiences: Archetypal sketches, structural diagrams, integrative exercises and techniques, and relational revelations of some of unity's deeper truths. I also share how I've put these gifts into practice as my work has unfolded over the past few decades.

My approach here is a-traditional and eclectic. It does not follow any particular lineage, regime, dogma, methodology, or schema. These stories draw upon inspiration from many varied sources, as do my life and this larger body of work.

I've decided to present these vignettes in an illustrated "short story" format, instead of as a comprehensively integrated treatise on the unity theme—leaving that complexity to the *Global Awakening* series as a whole. My hope is that these brief, faceted glimpses will work well with our contemporary time constraints and attention spans, and that their diversity of content will effectively engage an equally diverse audience.

My life's teaching has revealed many ways to experience "becoming the mountain" of oneness—from individual internal integration and uniting with the world at large, to uniting various aspects of our world. I've found that each way has an important role to play, especially when we turn our attention toward uniting to heal ourselves and our world.

The deepest essence of the unity vision that drives the broader and powerful healing impulse for the complete series, and how it came to be seated in me, is in these pages.

The implicit and, I hope, inspiring message to readers of this and other volumes is simply this: The insights, inspirations, and innovations that I share here can come to you as easily as they've come to me.

Other installments of the *Global Awakening* series delve in far greater detail into various specific aspects of this marvelous oneness vision—from how we're internally and globally united to how we can together heal our world. To learn more about the complete *Global Awakening* series and these more detailed aspects, please visit www.primamundi.com.

THOUGHT AND FEELING   CULTIVATE WISDOM (HEALTHY MIND)   VISIONARY INSPIRATION

RATIONAL   TAME   CELEBRATE   WILD   VISIONARY

WATER (CREATION)
(CARING/TASTE, PURIFICATION;
GRIEF (TEARS), RELEASE, HEALING
CULTIVATE COMPASSION)

ETHER: TOUCH/LIVING AN INTEGRAL LIFE
SELF/INNER/MEDITATION/PRAYER; THE UNCARVED BLOCK
PRISTINE EVER-BECOMING/ SOURCE

FIRE (DESTRUCTION)
... TION/VISIONS
... , PROVOKE, COMMUNICATE
... HAPPINESS & JOY

ONE

GLIMPSES

*Identifying and honoring all these ingredients,
leaving nothing out, helps us
see and live our life as a wholeness . . . .*

*You may find that you can actually become aware of
(one with) your very awareness itself. Now you can
experience your awareness as the incredible gift that it
is—your true essence—pristine, shining, immutable. It's
what we could call, "the present in the moment."*

*What really matters is
whether love is present or not.*

*Every sort of person was in
that room, and we were all
there working happily and
enthusiastically together.*

*An estimated 400,000+ participants
gathered in Manhattan, New York on
Sunday, September 21st, 2014 for the
largest climate change demonstration
in history. They were joined in
solidarity by hundreds of thousands
of other participants in approximately
2460 gatherings in 159 countries
around the world . . . .*

*Remember that your hand, like every part
of your physiology, is constructed with atoms
that were once part of other human beings
during every stage of human development in
the past seven million years.*

The possibility of awakening to and living in oneness belongs to each of us . . . .

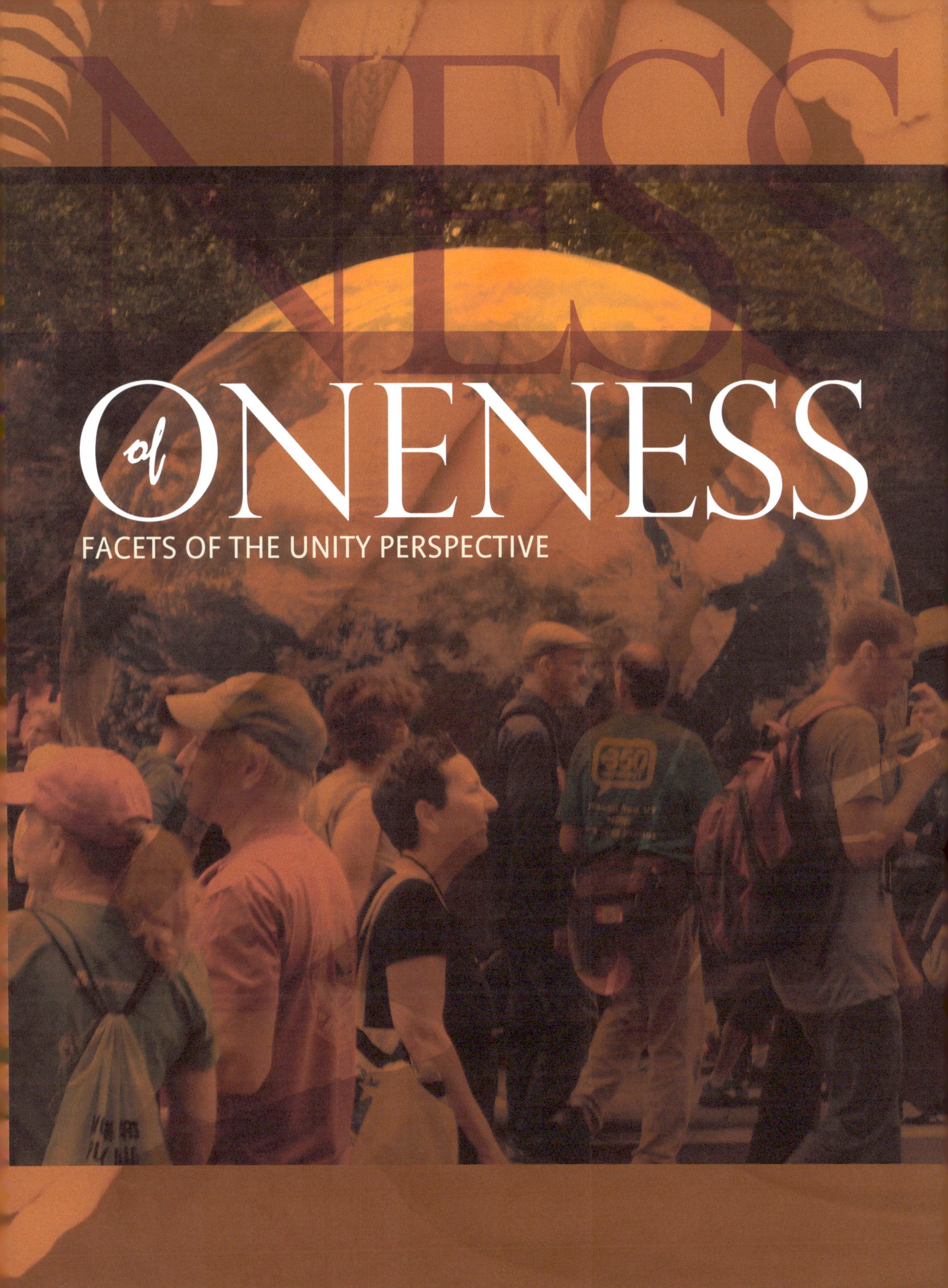

INTRODUCTION

I've been at sustainability long enough to have seen many of its advocates come and go. I still fondly remember the feeling I had the very first time I took a box of cans and bottles to be recycled, nearly a half-century ago. I was in grade-school, and our school's first recycling program had just been created by our newly formed "Ecology Club."

That sunny Saturday afternoon, I carried my box into a basement room full of activity and clamor—folks barking directions and others following them, clanging bottles and cans, etc., as the new recycling bins gradually filled. What struck me strongly then was that every sort of person was in that room, and we were all there working happily and enthusiastically together. It didn't matter who you were, or where you came from, or what you believed in. What mattered was that you were there and you were participating. This became my lasting "We're all in this TOGETHER!" imprint, the powerful sense that we were all uniting in action to make our world a better place. It was a very special time for me, one of those rare moments that can profoundly change your life trajectory, even though it may take you decades to notice.

We're all in this TOGETHER!

(Unfortunately, I've noticed that this feeling is sadly missing from many Earth-healing efforts lately. It's something we need to recapture soon in a big way if we're to survive.)

I followed the Earth-healing thread in many different ways through the following years. Sometimes it was in the foreground, sometimes in the background of my life.

Before spending the past twenty years in deep retreat, I was a successful architect and educator. I was an activist in various civic board positions, most notably as a founding board member of EcoVillage at Ithaca, an internationally acclaimed example of sustainable design. I earned valuable professional credentials, and eventually discovered that they were not what was really important to me. I felt a strong calling, what you might call an inescapable urge, for something deeper.

This prompted a move to a pristine, high-alpine/desert environment where I could empty myself of the accumulated flotsam and jetsam of thirty years of professional and spiritual development. It was a lengthy, somewhat painful, and extremely cathartic process that I would not have missed for anything—a time of minimal mental content, no professional accolades, and lots of peace and quiet.

Challenger Peak in Crestone, Colorado

I made pilgrimages to Europe, India, South America, Easter Island, Hawaii, and the sacred sites and pueblos of the American Southwest. I was privileged to hear teachings from many of the world's great spiritual traditions. I met and engaged with many remarkable spiritual elders and deepened connections with old spiritual friends and mentors. I went on week-long vision quests and retreats in our wondrous Sangre de Cristo (Blood of Christ) range of the Rockies, and connected deeply with the Power behind nature there.

What emerged for me during this time was a profound inner communion with All. That experience[5] became the wellspring of my daily life, and the impetus to share what I found became the driving inspiration to write this series.

These years also extended the reach of a fundamental aspect of my personality: the unified Renaissance man/can-do mechanic. Conceptual theorists generally express their ideas intelligently, but often have no experience with real-life implementation. They usually don't get their hands down into the dirt of actual situations where they can put their theories to the test. Conversely, many can-do/hands-on people lack the big picture, have no theoretical overview to guide their work. Fortunately, the capacity for both theoretical thinking and hands-on application are hard-wired into my makeup. My years of retreat became a time when I turned the abstract theories and lofty truths of my life's spiritual and sustainability-oriented leitmotifs into practical, down-to-earth, experience-based certainties.

During these years, for example, Carol and I designed and built our beautiful sustainable homestead. In the process, we managed to reduce our real household carbon footprint to less than 15% that of the average American household, the lifestyle I abandoned in 1993.

Honoring these truths in how I live has become one of my most demanding challenges. I'm not doing it to advance my career, gain a following, or make money. I'm doing it, plainly and simply, to support the vital change underway at this pivotal moment in time—aspiring to help counter the real and growing threat to life here as we know it.

I see the essential nature of global awakening as primarily a unifying, re-integrative phenomenon. I use terms like "unity," "oneness," and "wholeness" to express my views about it.

What do I mean by "Unity"?

For me, cultivating a stronger sense of unity in myself and in relation to others and the natural world is a powerful catalyst for positive personal, societal, and indeed global awakening transformation—particularly in relation to pressing planetary challenges such as global climate change. "Unity" is the most available, flexible term I know whose meaning everyone everywhere immediately understands. Something deep inside us responds to this word—as if we instinctively know what it means, whether we can articulate it clearly or not. And that's why I want to use it as broadly as possible, in a variety of ways, contexts, and circumstances.

For instance, I often use "unity" as a synonym for the wholeness that exists when any part of a whole clearly can't exist independently of the other parts, or of the whole itself. These parts (planets in a solar system, atoms in a body, particles in an atom, elements of Nature, etc.) can be said to "live in unity" with all the other constituent parts and with the larger whole.

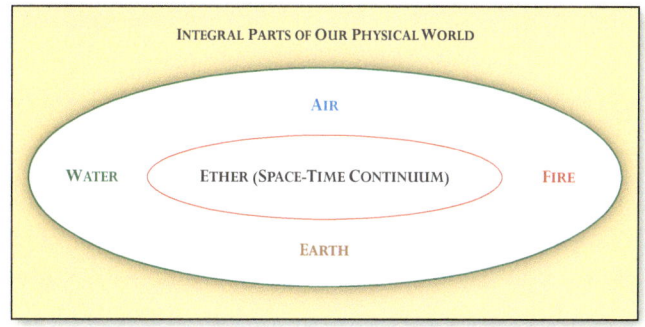

A diagram depicting the relationship of parts making up a larger whole.

If something can't exist as it does without the existence of something else, it is part of a larger system that includes, at a minimum, both of these parts. Webster's Unabridged Dictionary defines unity as "the quality or state of being or consisting of one: oneness," and as a "totality of related parts."

If we apply such interpretations to our shared human situation, we can speak of the underlying or fundamental unity of all things in this way: We each individually (you and I) cannot exist as we do without the prior existence of humanity, which cannot exist as it does without the prior existence of earthly life, which cannot exist without the prior existence of the Earth, which cannot exist without gravity and all the large celestial forms (like the Sun) that brought it forth, all of which couldn't exist without the prior existence of universal laws that apply to elementary particles, atoms, and so on. Ultimately, we'll see that all these forms cannot exist without the Big Bang, which couldn't exist without its Originating Power (known by many Names)—which is, of course, the broadest "totality of related parts."

Various philosophic, religious, or spiritual traditions prefer other terms for this inclusive sort of understanding: terms such as "interdependence," or "the nondual state," or Thich Nhat Hanh's "inter-being," or Ken Wilber's integral term "Holarchy," or Ervin Laszlo's "A-field," or philosophy's "logos," or "the Great Web of Life," "the Great Chain of Being,""the Wholeness of Universal Being," and so on. I prefer "unity" in large part because its simplicity easily includes all these more specialized interpretations. Using it broadly, gathering many different possible connotations together under its canopy, it has a better chance of making such understandings more comprehensible and accessible.

Our global awakening will require a full-fledged heroic collective effort. It's quite possible that it could indeed engender a new planetary mythology. Will our new epic end badly—with the continuing rape and ultimately the painful, desolate death of Gaia at the duplicitous hands of Big Oil, Big Coal, their political henchmen, and our own submissive complicity? Will humanity find the key that unlocks an ancient portal, a long-sealed door in our human consciousness, so that we can do the unprecedented on a global scale—unite with each other to heal our world? Indeed, will our new epic have a happier ending, with Gaia enthroned as she deserves, honored by our love, respect, and reverence as the giver of life, the Great Mother of All? Only time and what the overwhelming majority of us do will decide her fate and our own.

The possibility of awakening to and living in oneness belongs to each of us ....

Of this much I'm certain: The possibility of awakening to and living in oneness belongs to each of us—strong and weak, rich and poor, brilliant and uneducated. It belongs equally to Warren Buffet and Amy Goodman. It certainly is a reality for His Holiness the Dalai Lama and Nelson Mandela, who have consistently demonstrated its compassion in action, as have inspired change leaders like Emmanuel Ofosu Yeboah of Ghana or Olympic skating gold-medalist Johann Olav Koss, whose efforts have so greatly helped the Ghanyan disabled and disadvantaged children all around the world. It also clearly belonged to "Tank Man," the man walking home from work one day during the 1989 Tiananmen Square crisis, when, in an act of unpremeditated heroism, he stood in front of a column of tanks and refused to budge.

Our children, as the famous old spiritual goes, have the whole wide world in their hands, both its present and future. I hope that awakening into the unity perspective will help them hold it with intelligent compassion and full awareness of its precious worth.

ONE

LITTLE LEE'S *"MY CLOUD THEORY"*

I can still remember the first time I ever experienced the "wild blue yonder," or realized that I was a part of something much, much bigger than me—even though it happened some fifty years ago. I was just a little lad, four or five years old, and I was lying on my back in the cut grass of our backyard in Baltimore where I grew up. As I watched the clouds move quickly through the sky on a windy, early fall afternoon, I believed that they were *my* clouds, and that if I lay there and watched them long enough, the really beautiful ones that I loved the best would come back around again. I watched and watched and watched, but sadly, of course, they never did.

As it grew dark, my Mother called me in for supper—several times. Finally, my Dad had to come out and bring me in. I was crying my head off because *my* clouds had deserted me forever. My Dad kindly explained to me that they weren't mine to begin with and that they always changed, so I would never ever see the same cloud twice. So much for little Lee Temple's exceptional "my cloud" theory. I guess I never looked at clouds the same way again, and the world became a much more curious place for me from that day forward.

Many of us remember similar events—experiences that were simple enough on the surface, yet somehow managed to spark our imagination about the universe, make us aware of its vastness and mysteries and, if we were lucky, even make us feel a part of that mystery. Do you remember when it happened for you?

This adjustment in our outlook however, helps set the stage for a more profound understanding of the true equality of all things to enter.

As the fate of the dinosaurs and myriad other physical forms that have come and gone throughout the history of the cosmos attests, the universe can easily exist with or without us or our attempts to make sense of things. However, humanity and all its strivings and accomplishments would not exist without Originating Power, its Big Bang origination event, and the universe that unfolded as a result. To some, diminishing the value and pre-eminence of human power, knowledge, and wisdom like this may be unsettling, just as the debunking of little Lee Temple's exceptional "my cloud theory" was to me as a child. This adjustment in our outlook however, helps set the stage for a more profound understanding of the true equality of all things to enter. This kind of change must be at the heart of any deep, societal transformation that replaces our separative, self-aggrandizing, me-first tendencies with the shift into unity consciousness that we desire.

Our Earth- and human-oriented concerns, however sacred, are an integral yet minute (some might even say peripheral) part of an immense universe, the magnitude of which we glimpse more and more each day. Cosmologically, it seems the more we learn, the smaller we get in the overall scheme of things, yet the humility that comes as our awareness of this grows can actually help unveil our innate oneness with this much larger whole.

TWO

As I grew older and left "my" clouds behind, I soon developed an avid fascination and deep love for some of the great mysteries of our physical world, duly represented in our immediate neighborhood. From my perch in my treehouse in our backyard's big linden tree, I pondered many important questions. Like the time I learned that if I drove a tenpenny nail into a block of wood, wrapped a wire around it, and attached the ends of the wire to a battery, I could magnetize the nail. As I recall, this was the same year I received a microscope for Christmas (followed shortly after by a chemistry set, etc.). Soon the little scientist was on his way, discovering great things about our world— and luckily for me and my family, I didn't blow up our house in the process! I can still remember my excitement and awe the very first time I saw some of what I call "the little beings"—when I first looked at a hair, my blood, a cell, a bug, etc., under the microscope. Wow!

My worldly questions just kept getting bigger and bigger. Once my Mom took me to meet a very special lady—the famous sculptor, author, and Plotinus scholar Grace Turnbull. She lived just a few doors away from us in her very magical home/studio, complete with her own hand-carved totem poles defining its corners. For the visit, I wore my best cowboy outfit, including my dude hat, bandana, and toy six-shooter. When we met, she promptly and permanently took the gun away, vividly instructing me that violence was something never, ever to be celebrated. I was shocked that this petite elderly lady could do such a thing to me, but by doing so, she made certain that I learned an important lesson that day. It was my first exposure to such a genuine, critical intelligence, and one I've clearly never forgotten!

In my early teens, I had several mystical experiences involving grand old trees in our neighborhood. These experiences marked the first time I experienced a sort of unity in Nature. Back then I had no such name for my mystical nature or idea of what the experiences were. I just knew they made me feel full and happy. In the summer when school was out, I would take walks or bike rides in the afternoon around our block and neighborhood. Sometimes I'd become completely mesmerized by the way the light and the wind flowed through the leaves in the trees. The aliveness and vibrancy of their shimmering movement touched me deeply.

Around that time we had to make a demonstration of a natural event/phenomenon in biology class. I had been helping my Mom do some gardening in our yard and, along with some weeding, she had me dig up a little oak seedling that had just been birthed in one of the rosebeds.

When I dug it up, the little acorn that had birthed it was still clearly visible, identifiable, and firmly attached to the roots and shoot emerging from its shell just a few inches under ground. I was so inspired by and curious about the connection and relationships between the itty bitty acorn, the gestative soil, water, and sunlight and the tender, delicate little growing tree, that I decided right then and there to use it for my science project. This was a wise decision, because it ended up being a smash hit in class.

From then on, when I took my walks through the neighborhood and saw those huge, magnificent trees vibrating their aliveness in the summer breeze, I always remembered what a little digging in our garden taught me—that, just as I had seen with the little acorn and its seedling, it was only thanks to the miraculous power of natural survival and the unifying laws of nature that their own founding seeds made all these mighty beings possible.

THREE

A VISIONARY EMERGES

In my later teens my own personal galaxy began to burst wide open. Like most adolescent boys, I developed an intense interest in girls. It blended with an avid fascination for surfing and playing the bass guitar with friends in a high school rock band.

In those days, I learned what it meant to be part of a larger, unified team for the first time. My uncle, who also happened to be my junior varsity lacrosse coach, apparently decided that I could make a decent goalie (even though he soon became fond of teasingly calling me "sieve"). Reflecting back on it now, I realize that this simple assignment was one of the best opportunities of my early life. I learned many important life skills tending goal on his team.

First, you really learn quick reflexes when the most talented players on your and other schools' teams are swiftly whipping heavy lacrosse balls at you on a regular basis! Second and more importantly, the goaltender in lacrosse is responsible for marshalling the defense, especially by communicating ball position to the other defensive team players who are all busy tracking their opponents' every move. This important responsibility gave me a much stronger sense of presence and assertiveness, and all these factors helped me develop a greater sense of self-confidence on and off the field.

These were also the days whose weekend afternoons were filled with playing catch with my Dad in the backyard and attending local sporting events with him—especially the home games of the Baltimore Orioles and Colts, and the Johns Hopkins Blue Jays lacrosse team. He also taught me how to body surf during our summer holidays at the ocean, which brought forth a shared enthusiasm for the great gift of our life together.

One summer, I persuaded him to sell me an old '64 Ford Econoline van that was languishing on the lot of one of the movie theaters he managed, for the princely sum of $50. I fixed it up with paneling, curtains, etc., and it became my own little universe away from home. That was also the year I painted a 7' x 15' mural of a surfer catching a wave on one of the walls of my bedroom.

These were some of the first visible indications of the visionary artist/designer in me coming out to greet the world. I gradually began to deploy this wonderful gift of visionary imagination, first birthed in the

What really made a difference with any of them was whether love was present or not.

16th-century Italian baroque painter Michele Rocca's *"Angelica and Medoro"* depicts the two lovers on their honeymoon, carving each other's names into trees. Flying cupids with a torch allude to their burning love, while another spells out their names.

tentative questioning of my early childhood and then advanced in a sort of embryonic state during these teen years. It's the same faculty that, subsequently trained and refined, helped me to conceive and execute the *Global Awakening* series. It's a central aspect of my presence in this world.

The really interesting thing for me, as I look back at those years, was that, pretty much with all my interests at the time, what really made a difference with any of them was whether love was present or not. We can think of love as humanity's own form of gravity. When we love someone, something, someplace, they're pulled into our own personal galactic orbit, and he/she/it becomes a special part of us. The power of our love often transforms our relationship to them, and brings forth something that has never existed before. For me, as for many of us, it was these intense feelings of love that I would fatefully follow into the future. They subsequently opened a huge door for me several years later, but that's another story.

FOUR

LEARNING TO "LIKE WHAT IT DOES NOT LIKE"

Major challenges appeared in my personal landscape during the summer of my twenty-first year. I had decided to turn my emerging visionary, artistic/design talents toward practical applications, and by this time had become an architecture student in college. A design professor took a shine to me, brought me under his wing, and led me far beyond the normal educational confines of the architecture studio. His simple act of mentorship changed my life forever—and laid the foundation for you and me to meet through this book.

In the summer of 1977, while I worked with him on freelance architectural projects, he introduced me to the realms of philosophy and mysticism for the first time. Under his direction, I started reading books by spiritual teachers such as Gurdjieff, Ouspensky, and Krishnamurti, and this formally initiated a lifelong interest in philosophy and spirituality. Over time, I learned about the lives and teachings of many spiritual teachers and thinkers, some of whom I was fortunate enough to meet. I will always thank my professor for opening my mind and bringing these giant lights of humanity into my life.

In his own philosophic quest, he was associated with the Rochester Folk Art Guild (RFAG) in Middlesex, New York. This small, rural crafts community was founded and run on George Ivanovich Gurdjieff's philosophy of "the Work," an intensely focused method for encouraging spiritual growth and strengthening character. One day that summer, we drove to the community and he introduced me to the RFAG director, Mrs. Louise March, a powerful European woman who had served as Gurdjieff's secretary. A few days later, I learned that I was accepted for apprenticeship in one of the RFAG art studios.

I'll never forget my experiences there. All the people I met were interpreting Gurdjieff's work in exciting, captivating folk art expressed in varied media—painting, sculpture, woodworking, music, pottery, metal forging, and many others. They often worked in complete silence. Exposure to those great people and their deep sense of happiness and satisfaction was so different from my experiences in the rest of the world at that time—a world of increasing entrenched consumerism and collective fallout from the Vietnam War. My time at "the Guild," as we all called it, felt like a return to a romantic, bygone era. We woke early, meditated, prepared simple farm food, worked in the fields and in the studios, listened to contemplative music played by guild musicians, and attended teachings by Mrs. March. I thrived on all of it.

"You must learn to like what it does not like."

One morning after meditation, Mrs. March took me aside, held my hand, and gave me a brief instruction: "You must learn to like what it does not like." Then she smiled at me with her profoundly loving, luminous eyes and walked away. I understood the "it" in her sentence to mean my ego—that collection of self-identifying thoughts, memories, and beliefs that we all construct in our minds and usually allow to run the show. Our egos usually create and maintain a wide range of beliefs and accompanying judgments, from those we consider moral and life-supporting, to those that give rise to selfish, me-first tendencies.

Mrs. March had instructed me to like what my ego did not like. It took a while for that simple phrase to sink in and, of course, she didn't just leave it dangling there without any follow-up. She soon had me doing some backbreaking and menial jobs. She never verbally connected those assignments to the teaching, but they were definitely meant to drive home her message. For example, she asked me to destroy a large, old, stoutly built exterior stone fireplace with

a huge maul—a task that took me days by hand, but could have been accomplished in a few minutes with a backhoe. I had to do this noisy, sweaty, hard manual labor while other guild members worked quietly and peacefully in the flower garden close by.

I later realized that this was one of my first instructions about how to live with an understanding of the inherent unity of all things. As an up-and-coming architecture student, I had a rather inflated perception of myself and my abilities. Thus, at the start, I failed miserably at my assigned job. I was not at all good at it; my hands became painfully blistered and the experience showed me how weak and ineffectual I was in an area in which I had no experience.

I also had significant issues with the assignment because I thought the fireplace was beautiful and that destroying the handiwork of some other craftsman was sacrilegious. For all these reasons, I felt totally uncomfortable with, in fact, irritated by, the work. OK, I hated it. I swore under my breath and I could tell that the noise and brash violence of my activity was disturbing the peace and quiet—a quality that guild members prized above all others—of those working near me.

Then gradually a kind of magic alchemy began to take over. I got into the project more deeply. I became more united with it in my spirit and struggled less against it. By the time I was done, I was surprised to find that I had actually begun to like doing the thing that my ego had previously disliked. I was even a little sad when the work was completed, as I had actually done a good job after all.

From that time forward, I began to work with a clearer awareness of the potential unity that this simple teaching and its real-life context awoke in me. To be sure, I've had many more occasions in life to put liking what my ego does not like into practice.

For some, this teaching might reflect weakness, a lack of courage or the integrity to stand up for ourselves and what we believe is right. However, it takes great inner strength to tame the ego's powerful demands and commands. If we can find the patience and wisdom to quiet down the ego's knee-jerk reactions to contrary or foreign viewpoints or perceived threats, to watch and listen without passing instant judgment, we can give all people and situations a little more space. As we become more present with them, wonderful things can happen.

WARNING! CHANGE AHEAD!

use energy, modifications of the global and national economic systems that might initially seem threatening, weather-induced crises that produce huge human migrations and requests for aid, a demand for behavioral shifts at all levels to help save life on the planet and the human family as we know it. When they come, before we throw up our hands in denial, anger, or despair, let's give "like-what-it-does-not-like" a try.

Choosing to travel this path has brought me many beautiful gifts that would otherwise have passed me by. It has helped me live in a more complete unity with people, places, and events that have greatly enriched my life.

How does this story relate to producers and victims of global climate change? When I come across people who appear to be exploiting Earth's resources or inhabitants solely for personal profit, instead of indulging my ego's immediate response—some version of righteous indignation—I look for ways to turn that indignation into harmonious and active engagement with them. By choosing to like what my ego does not like, I can step away from my ego's judgments and find a way of walking in the other person's shoes. This brings my attention to our shared unity, the oneness that underlies all polarized beliefs and/or combative forces and people, including those at play in the moment. From this place, I have a far better chance of supporting a constructive resolution of any situation for the betterment of all.

I bring up this approach for us all to consider, and to inspire the many of us who will be asked, cajoled, prompted, coerced, or even forced to make changes that our me-first egos have no interest in making in the coming days, weeks, months, and years. Make no mistake, these changes are definitely coming: caps on carbon emissions that affect every way we

FIVE

THE POWER OF VOLUNTARILY EMBRACING GROWTH AND CHANGE

During the spring of 1979, I was deeply inspired by a visiting Swiss professor of architecture, Bernhard Hoesli. When I graduated from college later that year, I followed him to Switzerland, where I spent several years with fellow architects and educators that were part of his "tribe." I traveled and painted, photographed and sketched, hiked the mountains, read eminent philosophers as I sat in riverfront cafés, lived the bohemian life of Her-

mann Hesse, and saw the spiritual master Krishnamurti in Saanen. In short, I learned how to be, think and act on my own for the first time. Thanks to conversations with some especially gifted colleagues, I dove into the ancient, organic relationships among humanity, building, and the Earth that we can still find if we look deeply enough.

Then, with these valuable gifts in tow, I returned to the US in 1981 to teach architecture and urban design at

Luzern's beautiful Ruesse riverfront

several leading East Coast universities. I also met a woman, soon to be my wife, who introduced me to a group in Valois, New York, dedicated to the life and teachings of American philosopher Anthony Damiani (Wisdom's Goldenrod Center for Philosophic Studies, Ltd.).

Folks there explored the nature of reality, based on varied approaches from many different spiritual and philosophic traditions. As part of my involvement, along with other members, I worked with the associated Paul Brunton Philosophic Foundation, helping to compile and publish the writings of Paul Brunton, Anthony Damiani's primary teacher. This time provided me with a powerful, multi-faceted introduction to the experience of the *sangha*, or spiritual group, in a setting that honored numerous teachings. It also gave me my first exposure to the life and work of spiritual leaders and writers such as Plotinus, Ramana Maharshi, and Paul Brunton.

Concurrent with these expansive studies, my civic life during the 1980s and early '90s morphed in a way that could best be described as "three cats in a bag." By 1985, I was a licensed architect, a developer, and an educator. Professionally well known and successful, I was earning a good income.

This sounds excellent on the surface, but things were not so rosy inside. I was literally killing myself from overwork and living a compartmentalized, me-first existence, just like most people trying to advance their careers. Preoccupied with my life and work, I gave no thought to the origin of my food. I polluted and stressed my constitution with coffee and cigarettes and commuted two hours a day back and forth to work. I juggled teaching and my architectural practice; put on architect, developer, and professor hats as needed; and rarely took time to smell the roses. The early quiet interest I had in Earth and the natural world was completely buried beneath my driving ambition to get ahead. Sound familiar?

However, with the help of my wife and others, I also experienced seeds of a new awakening. Step by step, I became more conscious of the lifestyle choices I was making each day. I opened to the realm of intimacy in personal relations, through seminars and the work of people such as Stephen Levine, Joan Borysenko, Elizabeth Kubler-Ross, etc. I was also introduced to the Tibetan Buddhist movement, most

especially the presence and teachings of His Holiness the Dalai Lama.

When I look back, I'm amazed that, as I went through it all, I learned and changed so much. I can see now how

The author presenting plans to His Holiness the Dalai Lama at Wisdom's Goldenrod Center in Valois, New York, April 1991

my transformation and growth were inextricably linked to all the people and circumstances I encountered, which ultimately led me to the understanding of unity that I share here.

We live in an historical moment when our survival demands that we consciously transform the life-negative aspects of our individual and collective human nature, so that we feel and think and act from the truth of our oneness. This is not a conclusion drawn by radicals or malcontents on the fringes of society, but the clear truth determined broadly and convincingly by science and spirituality alike. This is the time when humanity as a whole and the governments that represent us must stop dodging this truth and connect the dots between me-first behavior and the damage it causes to life on all levels—personal and collective well-being and health, pollution and overuse of Earth's gifts (including the air we breathe, the water we drink, and the soil that turns seeds into food), climatic havoc and its decimating effects on every facet of life on Earth. This is the time for change from the inside out, before we reach that tipping point, which could even now be upon us. Truly, united we stand, divided we fall.

I accepted the truth and <u>voluntarily</u> made the necessary changes to avert a personal disaster. Just as we must do today, individually and collectively.

My own life experience demonstrates that we can *all* make this heroic movement; we require no special traits to accomplish it. Though in my search for fame and fortune, my me-first, more/bigger/better/faster ego was in full control of my life, I was somehow able to see that I was heading toward an abruptly terminal cliff. Once I understood this, I accepted the truth and *voluntarily* made the necessary changes to avert a personal disaster. Just as we must do today, individually and collectively, I stopped letting my me-first agenda run my life so that I could heal myself and others. I sacrificed some things along the way, but what I gained was far richer. I'll share some of these personal gems with you a little later.

SIX

Now let's turn our attention to more psycho/spiritual ways of apprehending and experiencing unity. All creation embodies or expresses the same infinite Source. Thus, for many human wisdom traditions, the way to enter into unity with the rest of creation is to become one with that Source. Since it is not only present within us as our own essential nature, but also in everything in the manifest universe, this should be and is a natural process. It's simply a matter of finding the most suitable route home, the path to oneness with which we resonate most strongly. For some of us the way home begins with self-knowledge and self-healing.

In the early 1990s, I started spending more and more time at the Wisdom's Goldenrod Center, and eventually formed a relationship with another member. Working closely together on getting the landscape in shape for a visit from His Holiness the Dalai Lama in 1991, we fell in love and started a life together.

This new relationship brought a strong and much needed connection to the Earth into my life, as my new paramour was an avid and exceptionally gifted gardener. She also brought me to an Oklahoma healer, Gail Lang, who changed both our lives forever. Gail introduced us to a holistic way of life that lifted me out of the ethical jam into which I'd gotten myself professionally. She helped me begin to work consciously with the "three cats in a bag" which had composed my life and to get them talking to each other again. In addition to learning and benefiting from Gail, I used dream and energy work based on Carl Jung's active imagination method (which I learned from a gifted psychotherapist named Barbara Prudhomme) as well as shamanic and spiritual practices. I was eventually successful at untangling those three cats and giving them the integration, clarity, and direction I needed.[6]

Through this work, I recognized in myself a visionary quality which loves to bring new things into being, and also a deep love of and concern for the Earth. Previously, these two things were separate and actually working at cross purposes. Working with Gail helped me find the courage to do something about this divided internal state. Through active imagination and visioning exercises, primarily oriented around asking myself how I wanted my life to feel, rather than what I thought about it, I was able to unify these two elements.

Gail also introduced me to Bill Bauman, the first person to encourage me strongly to love and assimilate all parts of my being. His love and support helped give me the strength to embrace my shadow or hid-

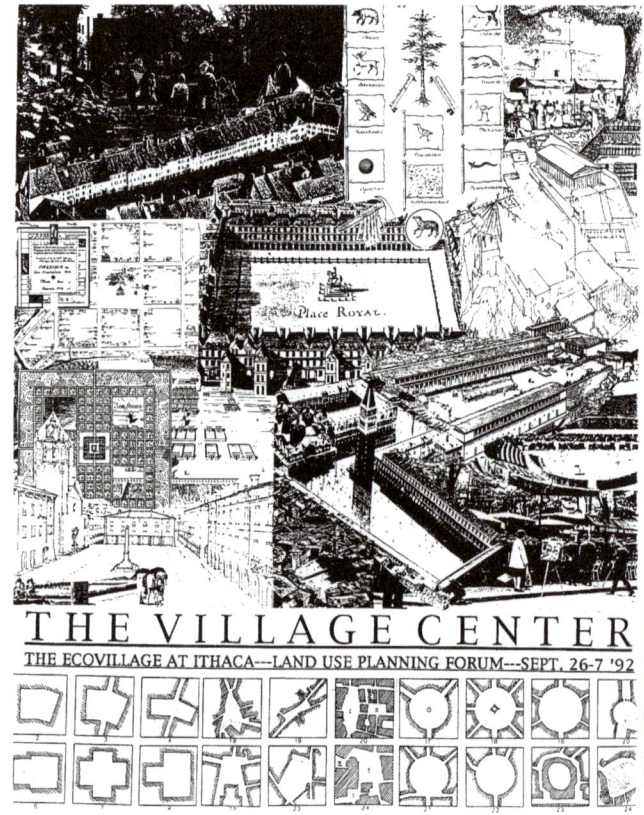

A collage I made for a village center planning document.

den side. I met Bill in late 1991 in the Washington, DC, area. A man with remarkable gifts, both wide and deep, Bill was also the first person to see clearly and honor all my gifts.

Coincidentally, at that time, a new idea for development was being floated in the Ithaca area. Led by Joan Bokaer, a small group was gathering to begin discussing the creation of a new form of ecological development, called an "ecovillage." It was based on a concept, first fleshed out in Eurpoean co-housing circles, that integrates living with dining and working in order to be less energy-intensive than the traditional suburban development models, where all these functions are separate. The ecovillage would take this several steps further and add agriculture and food storage, recreation, energy generation, and an educational component.

When I first learned about this project, my heart leaped, because I recognized instinctively that I could finally connect visionary development with environmental concerns. I volunteered immediately and was brought onto the initial planning team that eventually became the Founding Board of Directors. I did some early work on the educational center (their program was very much like that of my architectural thesis years earlier) that has just recently been implemented there. At that time, I also offered masterplanning ideas for the Wisdom's Goldenrod group and was able to merge development ideas with the spirit in that work.

One tool that was extremely helpful to me during this transformative process was the collage sensibility, which I learned while studying architecture at Cornell when the eminent architectural historian, Colin Rowe, was teaching there. He and Fred Koetter had recently written the book, *Collage City*. In addition, during my time in Switzerland, I worked with the Hoesli team, all of whom were excellent collage artists. I still remember Bernhard sharing one of the collage method's

It's not what it is; it's what you make of it that really matters.

important lessons with me during an exercise: "It's not what it is; it's what you *make of it* that really matters." Whatever we find, we can make beautiful. He also admonished me to "keep your feet in the mud and your head in the stars," like the lotus flower or the mighty sequoia tree.

The collage approach honors rather than judges everyone and every element of experience for what it can contribute. Employing this perspective helped me to take Mrs. March's like-what-it-does-not-like teaching and concretize it in my life. It is the sensibility of the fox, which uses all appropriate means to achieve an end, as opposed to the hedgehog, who only knows "my way or the highway."

I brought this unifying worldview into my turbulent life and professional situation during the early '90s to help me harmonize the elements I was integrating. My work with collage and studies at Wisdom's Goldenrod then led me to another breakthrough: I began to interpret the collage organization of the components of my being, life, and work through the lens of the *mandala*, a symmetrical form used in Buddhist and Vedic traditions to represent a deity, the home of a deity, or a map of the universe. Employing the mandala as a form of relationship diagram provided a way for me to identify and value all the absolute/relative parts of my life, and envision their harmonious union with each other and with Divinity every day. Doing so brought me more balance and equanimity.

I include a representation of such a diagram, what I call my Life Relationship Diagram, for your reference here. Building one of these can be a good way to remember the five fundamental elements (physical, mental, emotional, spiritual, and life cycle) of human nature and bring them into a more cohesive, consciously unified relationship. It also reminds us of the sanctity of every part of life. Identifying and honoring all these ingredients, leaving nothing out,

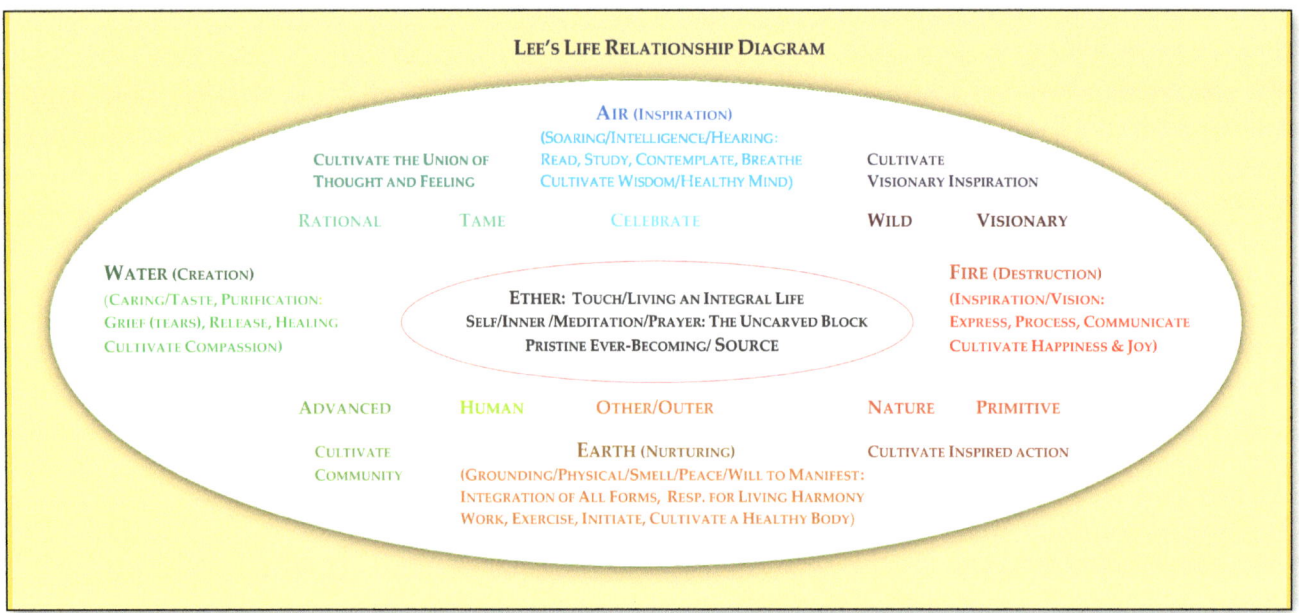

A diagram like this can help us to live with a greater sense of balance and harmony by creating a unity of opposites among the multi-dimensional elements in one's being and co-creation of the world.

helps us see and live our life as a wholeness, instead of as three (or more) cats in a bag!

It took me quite a while to get the hang of working with the different parts of my personality and bringing them into my conscious awareness. I used a combination of daily awareness exercises and psychological education and therapy to help the process along.[7] Some parts of ourselves are naturally harder than others to identify, accept, honor, and combine with the other parts. The challenging aspects are usually the ones that hide in the shadow recesses of our psyche, avoiding exposure to the bright light of day. These shadowy aspects and primal instincts are as important as those that are culturally correct. They deserve an equal seat at our internal unity table, since all are inter-related facets of our existence and expressions of our essence, just as all of the universe's many forms are manifestations of Originating Power.

If you decide to embark on a path of inner exploration and unification like this one, it may bring up disturbing issues and feelings. Therefore, having a guide who is trained in helping people go through such processes (e.g., a therapist or healer), and a good support system of friends will help you work truthfully and successfully. Don't be afraid when you find the parts of yourself that have been hiding. (This happens immediately when you try liking what it doesn't like.) We all have them. Don't be afraid to get to know them no matter how much pain, sadness, or anger they carry. Don't be afraid to find a good place for your shadowy parts, your animalness or whatever, at the table with your brighter elements. This process can be one of the hardest you'll ever face, but its successful completion brings the reward of integration and peace.

Identifying and honoring all these ingredients, leaving nothing out, helps us see and live our life as a wholeness . . . .

Why is it important to identify and integrate all the components of our inner life, including the uncomfortable ones? If we don't, for one, they will continue to influence our behavior and capacity for happiness and peace, usually not for the better. In addition, we may find ourselves projecting our unrecognized aspects onto others, especially those with whom we are either most familiar or most unfamiliar. Usually, we project onto people at home or work,

Sometimes projection becomes a group phenomenon. Group shadow projection is especially visible today in competing religious and political arenas. We tend to demonize others and turn them into the enemy. However, if we look inside, we find that the things that we don't like about them are often attributes of ourselves that we are unwilling or unable to admit we possess. For instance, if our buttons are continually pushed by having a boss who doesn't recognize the value of our work and always tries to make us do better, we may discover something inside ourselves that isn't satisfied with our daily performance.

When we stop projecting unresolved negativity onto others and own it, we can begin to see other people with more equanimity and with less of a good/bad charge. Seeing others clearly without projecting judgment is a big step towards seeing them as equal and, ultimately, as one with us.

I've practiced inclusiveness in writing the *Global Awakening* series, identifying all our important ancestors—the major aspects of the unfolding story of the universe and of earthly life—and then incorporating them into the narrative, so that all can be seen and honored for their role in creation and our experience. Working this way on ourselves is a creative, dynamic, and ongoing process in which we can engage any time, and which can have unexpected, tangible benefits.

For instance, in my case, I began to have lucid, powerful dreams. One night in the winter of 1992–93, I dreamed about having my hands in the vibrant red earth of the Southwest. It spoke to me vividly, and the strong, visceral feeling just wouldn't go away.

Finally, in the spring of '93, Carol and I took a trip out there. We visited many interesting places, but one spoke to us above all others: Crestone, Colorado. We both felt a powerful pull to Crestone and to one piece of property in particular. Trusting this emerging homing instinct, we purchased it two days after we first saw it. Within a month, we had packed up and moved there, even though we didn't know a soul in the town.

With this move, a new chapter opened in my life that involved heading out on my own more and more. A former student of mine came to work with me in the Sustainable Resource Center that I founded to advance my budding enthusiasm for sharing my progress and helping others live and build more sustainably. We did architectural consulting for clients, wrote a newsletter, attended conferences, and generally did our best to promote the newly emerging sustainablity worldview, in a town already inhabited by a small cadre of sustainable and alternative energy proponents.

In late '93, culminating a year of listening to our hearts, Carol and I finalized the design of the central elements of our homestead, in which we used lots of the ideas and integrative forms that were emerging for us then. As I reflect on this privotal time in my life, I see that we definitely seized the day, and took steps to boldly alter our lives. This period turned my ship onto the course that led me away from destruction and toward personal healing, the course that has led me directly to you.

The change involved in bringing me to this point was huge: I shifted from limited, fragmented, compartmentalized, and conventional to alternative, comprehensive, and holistic ways of being. I mention this here to help give you courage and strength for the shift that we're all about to make together. Steps like these are big, but as my example here shows, they don't have to be daunting. If you face them yourself, please remember: If Lee can do it, you can do it too!

SEVEN

EXPERIENCING OUR UNIFIED AWARENESS

I'd like to share an exercise that came to me about a decade ago. Some may wonder *why* it's so important to "enter the Now" to experience unity. I've found that it's necessary for a simple, mechanical reason. We have to be in the Now in order to be connected with or hold on to our very own personal, authentic, perceptual experience, our own direct *awareness* of the intrinsic oneness of all things. Fortunately, I've developed a few tricks over the years that can help us genuinely experience the present moment more easily.

Only by being fully present in each moment as it unfolds can you really get in touch with your own true-self awareness, which is your very own unity state. Watch it and follow it each moment, as all the contents of your senses are made into one experience there, in your awareness. You can access it only by being present in the Now moment. Experiencing this type of perceptual unity just doesn't happen any other way.

Let's first bring our attention to our conscious awareness—that great gift from our Innate Life Intelligence that has been crafted and developed through countless living species over billions of years. Normally, we spend most of our time with our own personal awareness pretty much plastered all over the contents of our experience. For example, if you were in a garden and looking at this beautiful peony, your conscious awareness would be *all* over it. You would smell its intoxicating fragrance and see the beauty of the light as it plays over the peony's reproductive organs and petals. If you're like me, its intoxicating presence might even give you an uncontrol-

> *Watch it and follow it each moment, as all the contents of your senses are made into one experience there, in your awareness . . . .*

lable urge to stuff the whole thing into your mouth and eat it—but let's not go there right now!

When your attention is given entirely to a great beauty such as this peony, you are not very aware of your awareness itself. Indeed, when we see or feel an attractive person, object, being, or phenomenon (food, art, meteors, etc.) we usually let it seduce us completely, just as this peony has done. Our awareness "goes out" to the object of its affection or attention so totally sometimes that we lose sight of everything else—even ourselves!

This sort of total immersion of our awareness and its focus is certainly one form of perceptual unity, a really great one indeed. It consumes a great deal of the energies of many an artist, scientist, athlete, hunter, soldier, religious practitioner, etc., and with great ef-

This beautiful peony in the Wingspread flower garden can symbolize the object of our awareness in the following experiment.

Pull it back!

Having first experienced the intrinsic unity of all things through your unified perceptual awareness experience, you can now become one with your unbounded source, the source of all creation, all perception, everything.

fects. However, it's not the type of unity experience I want to discuss here. Let's dig a little deeper by trying an experiment:

1. Choose something you love to experience, like this peony, and then put your usual attention on it. Try to notice that your awareness gets plastered all over the object(s) of your affection. You may get so involved with it, that you forget yourself, though you are the source of the awareness/attention.

2. Now, try to stay with the object of your attention, but *pull your attention's focus back* inside, so that you begin to see the object within the larger field of your experience. In the example of this lovely flower, you would begin to pay more attention to the other leaves and flowers around it, instead of just focusing on the big beautiful blossom itself.

3. You will now begin to see everything in the field of your awareness more equally: You will see the object (e.g., the peony) and the background of empty space, green leaves, and other flowers as one single, unified field of awareness. Architects like me have been known to call this phenomenon "figure/ground equality."

4. Now pull your awareness back even further, so that you begin to feel all your experiences in the entire expanded field of your awareness—including all the contents of your five senses (taste, touch, smell, sound, sight)—as one seamless phenomenon. This is what we could call our perceptually unified awareness.

5. Try to maintain your attention on this unified awareness for a little while. When your awareness is

Even further—pull it all the way back!

fully pulled back inside, and you are no longer fixated on any particular content, you can begin to see that all the elements of all the content of your experience (both figural, like the peony, and background, like the green leaves, other flowers, and the space between them) all exist there in a unified, equal state.

6. Shall we press our luck? Let's try one further pulling back of our awareness. When you get the feel of having your awareness pulled back enough to see all perceptual content as a unified whole, then try pulling back a little more. You may find that you can actually become aware of (one with) your very awareness itself. Now you can experience your awareness as the incredible gift that it is—your true essence—pristine, shining, immutable. It's what we could call, "the present in the moment."

This mightily pulled-back awareness stance has been called the "witness" position in some spiritual traditions. If you stay in that place of understanding that all you experience is really one thing at this level, and that it is your true self, your pristine awareness itself, that which you truly are at your very core—then you will have

seen and experienced the underlying unity of all things through your unified perceptual awareness and become conscious of this deepest, purest Source of your being.

In this fully pulled-back or merged awareness, all content dissolves into the truth of your limitless nature. Having first experienced the intrinsic unity of all things through your unified perceptual awareness experience, you can now become one with your unbounded source, the source of all creation, all perception, everything.

> *You may find that you can actually become aware of (one with) your very awareness itself. Now you can experience your awareness as the incredible gift that it is—your true essence—pristine, shining, immutable. It's what we could call, "the present in the moment."*

Congratulations!!! You've done it—you've experienced for yourself the perceptual unity of all things and your pristine Source Awareness!!! You've actually seen how everything you could possibly experience lies there within the unity of your own shining awareness. Way to go!!!

7. Now just **be** that awareness.

When you pull it all the way back, everything is included.

And remember, the singular awareness that you've just discovered is within you, *is* you. Lofty and distant as the possibility of your own self-realization may seem, you've already caught a glimpse of it in this simple exercise. It's really just a question of remembering who you already are. And if you still feel doubtful, remember that, as this experiment reveals, we can truly accomplish all great tasks like this, one step at a time.

EIGHT

THE *FEELING* OF UNITY

LOVE TO THE RESCUE!

Uncovering the wondrous wisdom and/or the factual content of this intrinsicly unified perceptual experience, however, doesn't get us all the way to unifying our species to help save itself and our world. For that step into compassionate unity, the one that will bind us with the state of being and fate of our world, we need to engage our primordial human emotions.

Even when we come to see all we experience as part of us, part of our unified awareness, we can still remain indifferent, detached, and aloof from it. Many people are like this. Their sense of unity is a lofty and even a rather detached one: "live or die, who cares, what does it matter, it's all part of The Plan." And at the perceptual level, they're right. But our lives are composed of far more than merely indifferent perceptions, and so this perceptual detachment doesn't remain intact or in the driver's seat when it comes to engaging the feeling of our essential oneness with all things.

When you are united with any person or being, place, or thing in feeling—for instance, by loving that certain someone, someplace, or something special— you instinctively care about his/her/its welfare. You take care of him/her/it as you take care of yourself, because in a certain way that person, place, or thing is a profound part of you—joined to you by your love and affection. That's why people throughout our history and even prehistory have given their very lives for the sake of their beloved. It is a natural, primordial emotion that we clearly have inherited from our mammalian forebears (which thus also links us in unity with them). It is found in pre-human precursors such as the relation of the human mother/child relationship to that of the chimpanzee, the instinct that makes a mother hen defend her young against coyotes, etc.

Love is both a means to unity and its natural expression, especially as compassion. We see it at work in all parts of the human realm, deeply ingrained in our nature. It drives a soldier to sacrifice him- or herself to save a comrade, a friend or stranger to help someone in crisis, or a mother to give her all for her children. When Buddhists take the Boddhisatva vows, they commit their lives to work solely to liberate all beings from suffering. Helping to heal others (by extension, this includes the world) is the centerpiece of their spiritual practice. It celebrates and demonstrates the authentic love and compassion-based bonding at the heart of the *feeling* of our unity with all things.

Love is both a means to unity and its natural expression, especially as compassion. We see it at work in all parts of the human realm, deeply ingrained in our nature.

Those who live in such a state of unlimited Being usually find that compassion permeates their active relations with all that is. The world is, after all one's Self. One finds no separation anywhere. These good folks employ their strongly reintegrated knowing and feeling of oneness, together with their fiercely re-energized will to help others and the world, as they live their lives. Such people are often spontaneously motivated to make the kind of changes, locally and globally, that will transform human society and our planet.

When any of us engage in Earth-healing activities,

such as trying to reduce our ecological or carbon footprint, we are saying that we recognize that we are linked in a unity that includes consciousness, perception, feeling, and action with everything on Earth and with the state of the Earth herself. We have come to know Earth as ourselves. We are saying, at the levels of perception, feeling, and especially action, that we care about the fate of this world, and all it contains, as much as we care for ourselves. We are one with the creation. What happens to it, happens to us.

This deeper union is thus at the heart of all efforts to reduce our carbon footprints. The spiritual imperative to experience our underlying unity and the scientific/environmental urge to save the world are not separate. They are deeply rooted in each other. The realization of unity births intention and action based in wisdom, love, and compassion. The shift in our sense of self from individual awareness to whole-world awareness permeates all levels of our existence, including our minds, hearts, and bellies—the seat of our will to action, or what the Taoists call the *hara*.

Here's an example of such globally focused, Earth-healing, unified mind/heart/will effort in action. In late 1991, Carol and I attended a Kalachakra initiation, conducted by His Holiness the Dalai Lama, in Madison Square Garden. It was attended by thousands of participants, including many members of the Wisdom's Goldenrod community. The Dalai Lama has long been actively engaged with healing our world and inspiring others to do the same. At the conclu-

The spiritual imperative to experience our underlying unity and the scientific/ environmental urge to save the world are not separate. They are deeply rooted in each other.

sion of several days of teaching, he distributed small packets of honey locust tree seeds (pictured below) to everyone present. This gift was meant to remind us that his teachings on universal compassion and accepting vows to help all beings should not remain remote, mental niceties. They could and should take root through compassionate actions in our shared human/natural world. For instance, he urged us to plant the seeds so that they could become trees that help sequester carbon dioxide and produce life-supporting oxygen, thus helping counteract global warming.

I wish I could tell you that it's enough for us all just to wish or pray this global climate crisis away, or for some of us to sit back and watch while others do the heavy lifting. Unfortunately, it's not enough to be passively aware of the situation. Since our daily actions are part of the problem, we have to consciously change them. The best way to do that is to begin to live and act from the inspiration of allness rather than individual survival or convenience. Partial answers involving ideas with no action, compassion without action, or action without compassion are no longer sufficient.

His Holiness the Dalai Lama gave a packet of seeds like this to all participants at the conclusion of the Kalachakra intiation held in Madison Square Garden in 1991.

NINE

ELEMENTS OF A PERSONAL APPROACH

Now that we're acquainted with these human forms of perceptual and feeling unity, let's look at how we can encourage them to root more substantially in our daily lives. Just as there are many ways to experience intrinsic unity, there are also many ways to prolong or maintain our awareness of its presence once we've found it. I'll just mention a few here—ways that correspond roughly to the three main energy centers in the body (head/heart/hara) where our primary powers of knowing, feeling, and willing typically reside. First we'll look at the way of mindfulness (head–knowing), then the way of compassionate understanding (heart–feeling), and finally the way of loving kindness in action (hara–willing). I've found that these three aspects, when woven together as a whole, can be quite helpful for establishing, enhancing, and maintaining oneness in our lives.

LIVING WITH MINDFULNESS

Consciously creating a healthy, peaceful life for my family and myself makes a big difference in my ability to live in and act from unity. I benefit from meditating regularly and eating fresh, organically produced foods grown as locally as possible. I live in a beautiful, peaceful, natural environment and get plenty of healthy exercise. I keep my life as simple as possible, minimizing sources of stress and environmental challenges.

I recognize that I am fortunate in being able to make these conditions possible. However, I also try to live with awareness, presence, or mindfulness as much as possible, something any of us can do if this path inspires us. Though many other equally powerful approaches to inner fulfillment exist, I will share some detail about mindfulness here. It is a totally accessible practice and one that I personally find extremely useful.

Simply expressed, mindfulness is conscious awareness of all that constitutes the present moment: the details of outer and inner content, including physical sensations and emotions, our thoughts, and awareness itself. Every small gesture of daily life holds a wealth of experiences. For instance, when we cook, we can be mindful of the feel, color, smell, and texture of the food we prepare; the sharpness of the knife we use to slice the vegetables; the shape and feel of the pot in which we place the food; the heat of the cooking fire; the glow of the flame. When we walk, we can notice the feel of the wind and sun on our skin and the color, shape, and smell of the plants we pass. We can even be aware of the feel and look of the pen and paper as we write checks and pay bills. We can also be mindful of our mental commentary on all

Simply expressed, mindfulness is conscious awareness of all that constitutes the present moment: the details of outer and inner content, including physical sensations and emotions, our thoughts, and awareness itself.

these processes: I wish I didn't have so many bills to pay this month, I'm too tired to cook for my family, I wish we could eat out instead, What a magnificent day! etc.

As we pay close attention to both the inner and the outer content of each moment, we begin to recognize how much commentary our minds produce. From there, we can start simply to notice these thoughts and let them move through us without judgment or rejection. Soon we find that our thoughts are just thoughts. They do not necessarily reflect any absolute truth or have much to do with actual reality. We can let the thoughts go and participate in life without getting overshadowed by interpretations and judgments.

In truth, everything just is. Once we see that our experience is a function of our mental content, we can stop identifying with our judgments and perceptions and walk through life in simplicity. We can write the check without anger or worry and move on, seeing that all that is actually involved is a series of coordinated gestures.

Eventually, as our awareness of the fullness of the present moment grows, we can also begin to experience life's infinite essence in every detail and in the spaces or gaps between those details: the gap of Silence between the letters in a word, the notes in a musical piece, the leaves of a tree. Living in the present heightens the most precious gift (the present) that we will ever receive, the only one we all share without exception—existence!

I also find that practicing regular remembrance exercises is helpful on the journey to unity—whenever I remember. I find this to be an excellent way to build a bridge from the internal message, "I have to do this" or "I have to do that," to the truth of oneness. I like it because it requires no special training or education, just an ability to focus and to recognize the truth when I see it.

I look at whatever is in front of me at the given moment—during a heated debate with myself or others, during a lovely dinner at home, during a walk in nature with my dogs, sitting here writing this etc.—and I say to myself: "I am that." In other words, I consciously shift my attention from my thoughts and feelings to include all that lies within my awareness. I maintain that broad awareness and begin to feel it as a unity, a wholeness, which of course, it is.

As we saw in Chapter 7's awareness exercise, everything we can ever experience is part of our own Being; it's that simple. This is the unity that you can experience anytime or anywhere you choose to be conscious of it.

Then, I wait and see if the magical feeling of unification with whatever is in front of me begins to emerge. If it does, I stay with it. If it doesn't, I repeat the truth to myself again and again until it does come forth. Then I stay with that feeling, that presence, as long as possible.

You don't have to go out and get the experience of unity. Unity is already fully present, the truth of your being and all that is. Therefore, it is possible to have a sudden shift that allows you to experience that your awareness is not just yours anymore. You (your awareness) have become the All. The individuated, limited you disappears in the infinite.

When you simply empty yourself of thoughts of all kinds, you begin to feel more spacious, expansive, all-inclusive. Eventually you find the entire cosmos within yourself. As you regularly speak or think the words "I am That" and open to the infinite, you may begin to know this unbounded awareness as your true Self, your essential nature. At that point, you can begin to say to yourself, "I am This" or just "I am," meaning your unbounded consciousness. Then, you can just be it. If you experience this, try to rest in this place as long and as often as you can (when not operating heavy machinery); it is your true and everlasting home.

You can eventually begin to recognize this state of pure awareness in other beings: trees, children, neighbors, pets, even pesky insects. You will begin to feel an affinity with it and its expression as these other beings.

LIVING WITH COMPASSIONATE UNDERSTANDING

I find that consciously remembering to pay attention to what I've written in this series—we're here only because of all of the preceding events and processes, and we affect the lives of countless other beings each and every day—helps me stay sensitive to my environment, grateful for my existence, and humbly present in relation to all the other beings who share this world with me. I remember these things as often as I can, in all circumstances, even when I'm dreaming.

These revelations open a door to a new *understanding* of who we are, our real relationship with the world, and how we can come together to save it. When we understand that God's elementary particles populate everthing from cosmic dust to Copernicus, Venus to the Vatican, we can begin to treat everything, animate or inanimate, more equally. It's hard to do this consistently, especially when we have to eat, make a living, etc. The main thing is to do the best we can, starting right from where we are today, to live with a new, more compassionate reverence for all of God's Creation, even when we're eating part of it!

Compassionate understanding is what happens to us sometimes when there is a sudden spark of recognition—the "aha" moment that often occurs when our knowledge transcends itself and comes alive. We can *understand* that we need to unite to heal our world, for instance, because we have connected the dots of many individual pieces of knowledge about who we are, the consequences of our collective behavior, etc., and in so doing, open to the deeper realization of our oneness. Understanding can be the precursor to effectively putting such holistically unified knowledge to work. It also arises from a real sense of "figuring something out." For instance, when we see the connection between climate change and human activity, we understand that we must change our behavior to create a different end result.

There are several key aspects of compassionate understanding. We may find that we develop a stronger sense of *empathy*, for instance, with all other beings, places, and things. Another aspect is the unfoldment of *reverence*. Paul Wooodruff, in his book by that name, reveals essential truths about how reverence can enhance our understanding of unity.

LIVING WITH LOVING KINDNESS

How do we actually evolve and engage our understanding? This is a process that takes time and can't be done overnight, even as precious insights that can come quickly will enhance our understanding moving forward in our lives. We may find that insights gained from the journey through this book can begin to take hold. Mindfulness of such gifts and of the present moment help evolve our understanding in natural ways. Inevitably, the more we understand the essential unity of all things, the more caring and tolerant we are of them.

Our compassionate understanding, born of mindfulness practices and insight, can help bring forth new paths of loving action for ourselves and our world, now and in the future. Yet it can be hard to think of being kind to others if you don't feel good about yourself. Therefore, one crucial aspect of learning to live and lead with loving kindness is to feel content and peaceful.

One way to foster loving kindness, no matter from where we start on the internal/external happiness scale, is simply to cultivate generosity. When we feel unhappy, angry, fearful, or empty inside, we automatically put our own needs first. Sometimes those wants can be extreme, showing up as greed, desperation, or powerful cravings that damage ourselves, others, or the environment. We can consciously re-orient these feelings and behaviors by creating the habit of giving, whether we offer something material, care for or protect others, or share kindness and love. When we develop the habit of active generosity, it becomes an inward state and we can find ourselves extending generosity to ourselves as well.

We can also cultivate gratitude to heal feelings such as that life does not provide for us or that we ourselves are not "enough." We can begin to notice the ways in which life does indeed give us just what we need.

Finally, we can ask ourselves what makes us genuinely happy. What do we really require to be contented in our lives? What sorts of things are truly important for us and which are just fluff?

We will soon find that caring for others is good for the soul. As we grow fuller and more peaceful inside, compassion arises easily and naturally. We can practice caring for friends, acquaintances and even enemies—who are endowed with all the same faults and foibles, the same five fundamental elements of human nature, as we are. We can also develop an attitude of caring for the world as a whole.

Are you wary that apprehending and/or engaging the world with loving kindness will make you a doormat for others with aggressive or dysfunctional behavior? If you are, try to remember that compassion is an outgrowth of strength, not weakness.

We needn't be part of a spiritual group to practice loving kindness. Many non-spiritual organizations cultivate and practice it on behalf of the less fortunate all over the world, under the guise of disaster relief, feeding the hungry, clothing the poor, etc. Individual examples of practicing loving kindness abound as well.

TEN

THE TWO PRINCIPLES OF LIVING IN UNITY

1. EVERYTHING IN THE UNIVERSE SHARES A SACRED, COMMON ORIGIN AND EXISTENCE

To live from a unified consciousness is to recognize all parts of creation as your own Self. You value and honor all of it; it is all You. You live with love, reverence, and compassion for your five fundamental elements—your mental physical, emotional, spiritual, and life-cycle aspects, the entirety of your life. Honoring and experiencing all parts of the universe as your Self gives rise to a conscious recognition that everything in existence shares a common origin, is equally alive and equally sacred.

2. ENGAGE THE GOLDEN RULE: DO UNTO OTHERS AS YOU WOULD HAVE THEM DO UNTO YOU!

The Golden Rule is the vibrant heart of spiritual teachings and traditions the world over. What a simple way to live our intrinsic unity and equality! Though living the Golden Rule is a natural style of behavior once we live in unity, we can begin to practice it in the present moment. Because whatever we put our attention on grows, this behavior in and of itself is a practice to enhance oneness. It cultivates empathy and compassion and, ultimately, our ability to know others as our own Self, as intimate expressions of the oneness that is our true nature.

What does it mean to put the Golden Rule into practice in daily life? Imagine yourself in a schoolyard (or any business, government, community, or personal setting) during an unsupervised recess or break. When you think no one is looking, you see an opportunity to take advantage of somebody by asserting your su-

> # The Golden Rule
> *Treat Others*
> *As You Would Like*
> *To Be Treated*

perior strength, just to make yourself feel better about yourself or look better to others. If, however, you remember the Golden Rule, you will first ask yourself how *you* would feel if someone treated you that way. Would *you* like to experience the emotional or physical harm you are intending to inflict on the other person? Would you feel valued, respected, loved, or honored for who you are?

It's easy to see how both like-what-it-does-not-like and the Golden Rule can help move us towards unity. If you find yourself wanting to take advantage of somebody else, or see an opportunity to profit by their loss (e.g., signing up an unqualified person for a home mortgage, knowing full well that the loan will be catastrophic for them in the long run), ask yourself to like what it does not like. This means putting the other person on an equal or even superior footing to yourself. By honoring the other, you also honor yourself. By caring for their welfare at least as much as for your own, you enliven unity in your own awareness and set up a win-win situation for you both.

Life regularly presents countless opportunities for us to apply the Golden Rule. The more we practice it, the more we recognize them. I have a humorous one to share with you now.

ELEVEN

A DOG'S PROFOUND GOLDEN RULE TEACHING

Shih tzu sisters

If I wasn't fully present, I would get annoyed and pull on their leashes to hurry them up so that I could get back to my pressing (and to me, seemingly more important) human business.

I'm honored to live with two shih tzu dogs. This story takes place several years ago and is about one of them and her amazing intelligence and sense of humor. We live out in the country on the edge of the wilderness, so we can't let our dogs just run free outside to do their business, because the coyotes or the hawks could make an easy meal of them. While I do feel generous towards our fellow beings in the wild, I'm not that generous!

It used to be that when I took the dogs out to do their business, I'd have lots of things on my mind and, well, I would get kind of impatient if they were taking their time with sniffing all the smells. If I wasn't fully present, I would get annoyed and pull on their leashes to hurry them up so I could get back to my pressing (and to me, seemingly more important) human business.

Well one of these adorable beings, being the shrewd, wily little fox that she is, found a way of letting me know that she was not happy with this kind of non-equal, non-reverential treatment. Would any of us like to be hurried along by somebody else who thought they were more important than we are—especially if we're their special "baby"?

Whenever I went into the bathroom in our house and sat on the toilet, she knew she had the perfect opportunity to teach me the lesson that I needed to learn. She would barge in through the door and sit down right in front of me. She would then start to growl and prance around as if she was impatient with me.

If she could talk, I'm sure the verbal subtext to this treatment would have been something like: "Hey, c'mon Dad, I don't have time to wait around here while you do your business—hurry it up, will ya—pick up the pace—time's a wastin'!"

She did this so many times that I finally got the message. She was teaching me vividly about being present to what was really going on in front of me and living by the Golden Rule: Don't treat others in ways that you yourself would not like to be treated.

After receiving this lesson, with my tail between my legs, I made absolutely sure that my fine furry friends had all the time they needed to do their business outside, showing them the respect and honor befitting the regal queens they really are, and treating them as I indeed would also love to be treated.

You know, it really works! Now when she comes into the bathroom, she just lays there patiently waiting while I do my business. There is much greater peace and harmony in our household, in the realm of poop!

> *Applying the Golden*
> *Rule consistently will*
> *help us move from the*
> *passive to the active*
> *phase of healing*
> *our world.*

Applying the Golden Rule consistently will help us move from the passive to the active phase of healing our world. We need only acknowledge that action is indeed required, and develop a greater global human tolerance and celebration of each other and how differently we all live our lives, as we engage the potential of our collaborative collective cooperation.

Humanity is currently the self-reflective pinnacle of terrestrial life's evolutionary lineage, the incred-ibly fortunate beneficiary of all that Originating Power has created over 13.8 billion years, with all the accomplishments and achievements outlined in my Global Awakening series and many other books. We have walked on the Moon, built instant global communications and information systems, written magnificent music and poetry, lifted millions of people out of poverty and disease, and solved some of the great mysteries of the universe. Having all this and so much more under our belts tells us that we are fully capable of any required change.

Please, in all you do each day, find the big picture and ultimately the essential oneness in which all your experience floats. You and all of us will benefit immensely from these small daily choices. Acquiring a new style of living can be challenging at first; but it will get easier and easier, and soon it will be second nature to you. Your new connection with all life will bring you an incomparable fullness and richness.

TWELVE

IT HAPPENED ONE DAY

Assistance in making this change is available in many forms and places, so don't let the naysayers within or without dissuade you. Great shifts can be made in the blink of an eye. On a 1998 trip to India, for example, an extraordinary and illuminating realization of unity came to me, seemingly out of the blue. It took place in the ashram of the great South Indian spiritual luminary, Ramana Maharshi. I was not expecting anything before it happened. I made a heartfelt, authentic offering of sorts, and opened myself to be present and to receive. In other words, I put myself into the Now.

A view of Ramana Maharshi's ashram on a clear day, with the sacred mountain, Arunachala, in the background. See the peacocks on the roof to the left. Together with monkeys, they were the most prevalent animals there.

LEE'S DIARY ENTRY FOR FEBRUARY 23, 1998, 8 AM, RAMANA MAHARSHI'S ASHRAM IN TIRUVANNAMALAI, INDIA.

Today I had a very special morning. I had done a bunch of laps (circumambulations of the Ramana statue in the large ashram Shrine Room) and had gotten quiet before breakfast. I had also been sitting by the window outside the Quiet Room, waiting to eat, and, just before breakfast, caught a fleeting glimpse of the peak of Arunachala, the sacred mountain which was usually enshrouded in fog or clouds in the mornings. After breakfast, I sat by the window again for quite a while, looking toward the dispensary and Arunachala and enjoying the beauty.

Unexpectedly, I found myself gathering up all the sadness of my life and my lack of total fulfillment and offering it all up to Ramana. Then I opened myself up completely to receive the positive aspects of the place and his being that were beyond my comprehension or consciousness—those things beyond my limited capacities to understand or otherwise take in.

Here, come with me. There is something I want you to see.

Soon I became peaceful and still, and felt a special presence. I was very silent, and got comfortable, and then Ramana came, appearing in a loin cloth and grey fuzzy hair and beard. He told me to let go of the body for a little while, to just let it be and rest. He told me that it wasn't going anywhere. He repeated that I should feel completely safe, let it go, sink into the warmth of my true Self and feel the oneness and the peace, the deep, inner peace of this state.

At this point, I felt deep peace and letting go and I felt the oneness of myself, my surroundings and all beings. Then I began to see a light, and Ramana said: "Here, come with me. There is something I want you to see." He took me to sit with him in brilliant white light, just the two of us, sitting together in this brilliant white light and stillness.

Then he said: "This is why you need to go beyond the body thought," answering a question that was on my mind when I did my laps before breakfast. Then he continued: "This is the pure light of your soul in union with All—the Brahman Atman—and it is always here for you. Now I will always remind you of this place, and if you get your mind quiet, you can come back here whenever you can or want to."

Ramana explained that I didn't need to be in the Quiet Room to have these experiences, but that I needed to have or go to the Quiet Room in me, in order to have them. My path was not so much about meditating, but about seeing the divine in everything—in nature, the beauty of the Earth, etc.

He also asked me if I could feel the peace and tranquility and how rejuvenating it was. I said and actually felt my answer as a "yes." Ramana disclosed that this was a reason why people sought to gain or attain this state of being, to bathe in this light: He said it was totally rejuvenating, "the eternal fountain of all life and youth," and that only a few moments bathing in this light would make one feel as rested as a whole night's sleep.

He further imparted that this light is the eternal Fire of Creation that burns in all living things. It is the direct connection between all things and me, literally responsible for the oneness of all things, present in all things. Ramana asked me if I could feel the interconnection, and I did. Then he said: "This is why people come here and why we try to attain self-knowledge. It brings us into a constant state of peace, rejuvenation and connection to all life. (Interesting that total withdrawal can actually bring about total connection to all of creation.)

The wall outside the Quiet Room where this experience took place. I was sitting against the wall near the window on the left.

All of this only took a few minutes' time. My heart was totally open and relaxed, at peace, as was my body. The white light was ever-renewing but always still and constant; as if in a cloud, but also clear and radiant. Finally, Ramana said that I could go back to the world. I briefly opened my eyes, then went back inside and thanked him. He replied, "So, now you see why we do this."

Of the many spiritual blessings that have graced my life over the past several decades, this is probably the best personal example I can offer of an instantaneous shift into unity consciousness. It was a pivotal moment that opened the door wide to my exploration and growing familiarity with oneness. It also planted the original seed for my *Global Awakening* series. So you see, whether you expect it or not, when you make a genuine effort just to show up with an open heart, amazing things really can happen.

THIRTEEN

TRANSFORMATIVE CHANGE—IT CAN HAPPEN TO YOU!

Change often feels daunting to people struggling just to make ends meet or to live another day, or who like to go about their business without demands and challenges from the "outside world." Yet change, particularly evolutionary change, is one of life's most compelling constants. Nothing in life, from the personal to the universal, stands still, whether we notice or not. For instance, it could be argued that yesterday's invention of the now indispensible cell phone and Internet birthed the more recent Arab Spring. The change in number and tenor of calls to action since I started writing this series ten years ago is also quite impressive. Increasing numbers of people are making more and louder demands for concrete shifts toward sustainable resource use, the restoration of natural systems, decreased fossil fuel consumption and less consumption in general today. Though we may have far to go, a profoundly evolutionary global transformation, involving *billions* of people engaged in *millions* of transformative actions, is definitely already in progress.

How can we as individuals participate more fully in this worldwide phenomenon? Most simply, acknowledge and respect the needs of others, and reduce our individual ecological/carbon footprint, especially by consuming less and conserving more. This involves both inner transformation and external actions. The latter in particular exist in abundance and a good number are widely accessible, affordable, and doable on numerous levels of human organization.

Although such transformative calls to action are happening more and more all around the world today, they are certainly not coordinated or harmoniously interrelated yet, by any means. So I believe it's quite important to establish

How can we as individuals participate more fully in this worldwide phenomenon? Most simply, acknowledge and respect the needs of others, and reduce our individual ecological/carbon footprint, especially by consuming less and conserving more.

a basic, holistic context for better understanding them. It's good to be mindful of several key points as we consider the many varied Earth-healing approaches.

First, each approach has a *convergent* mission with all others. In this global healing effort, we're all working towards (converging on) the same unified end of healing our world, even though it's still difficult for many of us to acknowledge this truth or to work cooperatively together—especially between currently antagonistic camps, and ultra-especially within the activist "heal-the-world" camp itself. Convergence consciousness is thus an essential aspect of mutual recognition *within* the global mix of multi-dimensional yet often specialized parts today.

Next, let's think in terms of phasing. Some of the strategies I discuss in the *Global Awakening* series are simple and can be implemented immediately. Others may have to wait until a first phase is achieved. Yet others are for the long term. Changes involving advances in technology or broad behavioral shifts definitely cannot be accomplished in a day!

For instance, when we consider vehicle emissions, strategies that involve the gas/battery hybridization of vehicles are already well underway. Witness the expanding popularity of Toyota's Prius, the new Chevy Volt, and other hybrid cars. In the near future, manufacturers will move to all-electric or electric/hydrogen cars; further out, they will move to all-hydrogen, after infrastructure transitions are put into place. Or, take humanity as a whole. We must move from robust growth to zero growth to negative growth in economics, population size, fossil-fuel use, and many other areas.

When considering possible next steps for humanity, the issue of moving toward some sort of partisan utopia inevitably crops up. Some people dismiss calls for action to heal the Earth as unnecessary, unrealistic, impractical, politically and/or financially manipulative, or even naïve. However, experience often shows that we can move toward an ideal solution using real world means that we can all understand. Moving from hybrid to electric to hydrogen-powered vehicles is a perfect example.

When we consider real world moves, as we are about to do, let's keep the ideal goal in mind. Let's also agree that for those of us who are not activists (which is usually most of us), it's okay to take small, universally doable steps, so no one is left behind. Let's also try to keep our ideal-seeking strategies and tactics up to date—focused on the best available methodologies at any given time.

Next, let's acknowledge that we humans often have conscious or unconscoius resistance to change, and sometimes even the best and brightest "change agents" among us get frustrated or angry with those who still don't acknowledge change's necessity today, and vice-versa. Let's make a point of managing our anger responsibly moving forward—by identifying, honoring, releasing, and understanding it, so that we can make our evolutionary, Earth-healing changes in healthier, more balanced ways. Anger only divides us, and it is extremely unlikely that everyone in the world will ever agree on anything! We need to take action based on the truth(s) we recognize and also to genuinely honor and celebrate others who do the same. To arrive at a healthier world in oneness, we need both to let go of our own anger and to see the other person's point of view, or to "like what it does not like." The last thing we want to do is create more polarization in world consciousness by digging in our heels and declaring that the only viable way forward is "my way or the highway."

And finally, in a related vein, among those people and groups who agree that action on climate change is necessary, let's avoid allowing our natural enthusiasm for constructive change to turn into missionary zeal. We don't want to become behavioral dictators or self-righteous proponents of the approach that has won our hearts and loyalty. To be successful, Earth-healing needs every person's contribution, however he

or she is inspired to make it. Our task is to find and follow our own inspiration and celebrate the same opportunity for others. Take two seemingly opposite possibilities for Earth healing—the methodical systems approach to carbon footprint reduction vs. the mystical approach of simply Being and allowing life to unfold from that infinite inner reality; either can be the best fit at different times of our lives.

Our Mother Earth and collective human consciousness are clearly going through all the painful, real-life traumas of a decades-long birthing experience. We're in global Earth-healing labor today! It is at times messy, conflicted and confusing, even frenzied, but we can appreciate every phase and factor as a part of the miraculous, unprecedented global transformational movement under way—just as it is! The

For some of us, engaging such change may feel like finally "coming home." For others, it may involve equally wonderful exploration of previously uncharted territory; and for yet others, it may be a delightful melding of these two extremes.

individual tree participates in the forest's unity while adding its unique form and character to the whole. So it is with each and every individual, community and nation: What each individual, community or nation carries, each contributes to shaping and enriching our world and her healing.

For some of us, engaging such change may feel like finally "coming home." For others, it may involve equally wonderful exploration of previously uncharted territory; and for yet others, it may be a delightful melding of these two extremes. As you envision your own participation in Earth-healing, don't feel obliged to adopt any particular approaches if they don't call out to or ring true personally for you. Simply let all the possibilities for change wash through and see if and how they call you to proceed. Acting from inner guidance, following your bliss, as Joseph Campbell expressed it, you will find yourself making the special gift to global human transformation for which you are specifically suited and designed.

What is most important is to stay in a place of love, inspiration, and effortless simplicity as you find your personal style of joining with the rest of humanity to heal our world. However, remember that to make a concrete difference, whether you are an idividual, family, school, business, community, state or nation; whether through meditation, activism, or anything in between, you still need to walk your talk: Just do it!

Our planet needs large numbers of people to make the evolutionary shift in attitude and behavior from me-first to world-first. Fortunately, the shift in consciousness towards oneness that this book and series so greatly emphasize will yield spontaneous behavioral changes: When you experience the world as your Self, you care for it accordingly. As you become more authenticly connected with all that is, you will find yourself both consciously and spontaneously working your planet-saving magic in the details of daily life, and in conjunction with efforts at all possible scales of activity—from investing in hybrid vehicles and recycling, to achieving caps on carbon emissions nationally and globally, to meditating. Now let's briefly investigate a few proven tools and techniques that effectively facilitate such comprehensive transformational change processes.

FOURTEEN

FOUR POWERFUL CHANGE-PROCESS TOOLS

As mentioned earlier, one way to implement transformative unity is to construct a diagram of relationships among the many parts and/or scales of human organization in which one is situated.

honor, accept, balance, and include them in a greater unification toward healing our world comprehensively.

While certainly not representative of every aspect of human endeavor, such diagrams nevertheless seek

THE RELATIONSHIP DIAGRAM

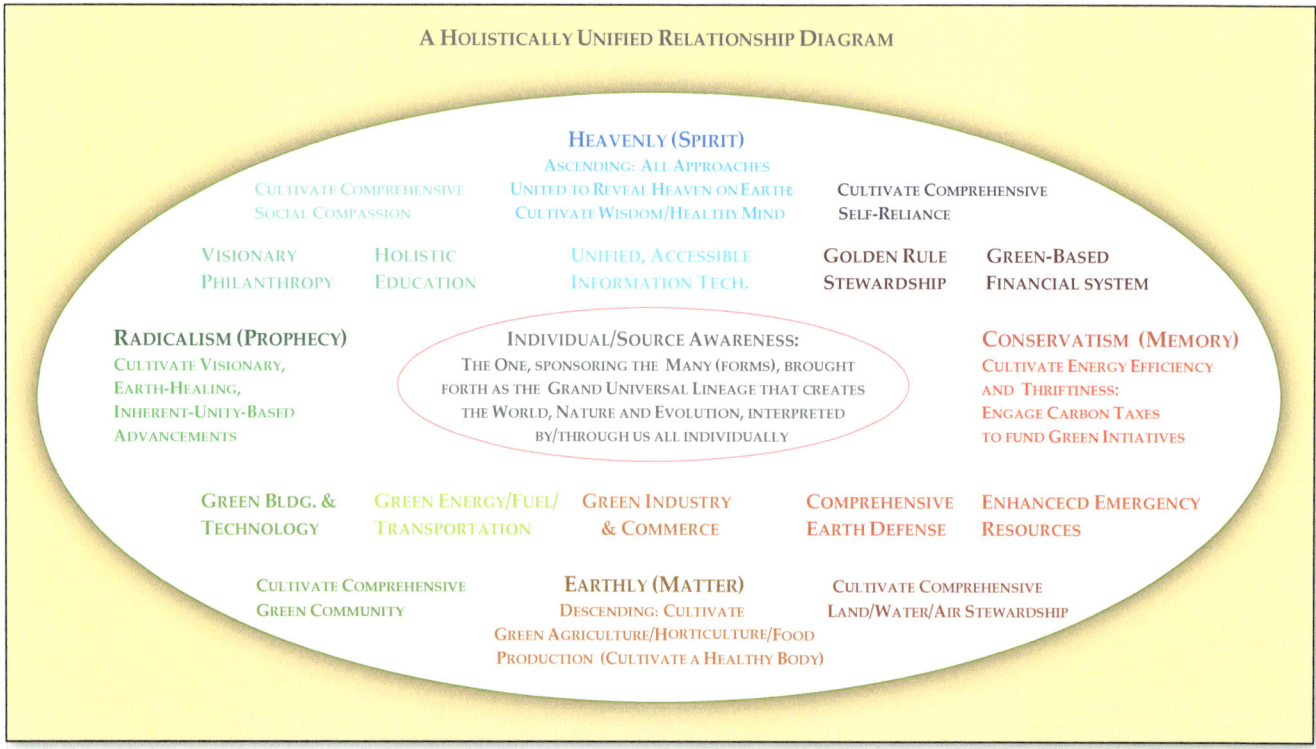

A HOLISTICALLY UNIFIED RELATIONSHIP DIAGRAM

This relationship diagram illustrates a comprehensive, integrative, honoring approach. It can apply to any scale of human entity-ship: individual, family, commercial/corporate, community, institutional, educational, state, national, multi-national, or global.

In this diagram, the periphery identifies key issues and directives related to or affected by global climate change. They all interrelate with each other and the Source of manifestation, just as all collective human qualities and endeavors (summarized in the five archetypal elements of human nature) relate directly to every individual. These qualities (we can call them ego forms) are integral parts of our individual and collective wholeness, and we thus need to identify,

to include the full spectrum of human paradigms and persuasions—political, religious, etc.—and reflect our total individual and collective relationship to Source and World.

THE WAY OF COUNCIL

Relationship diagrams such as this one can visually capture and communicate a unified structure and identify its components, but they can't actually create the relationship(s) or bind the various parties together in living, active union. For that, we need the help of other powerful tools. Another helpful technique for weaving human unity among component parts at various scales is *The Way of Council*. That is the name for the process I employed for my internal reunification (mentioned earlier), when it involved relationships with other people. I first came upon it in early EcoVillage meetings in the early '90s, and later found it quite helpful in a variety of settings.

As we get to know the pieces that compose our internal life and external worldly connections, we need tried and true ways to facilitate honest, respectful communication within and among those parts in order to engage in authentic, mutually respectful relationship. The Way of Council gives every part or party not only a seat at the table, but also a chance to speak its/their own truth without interruption, while all the other parties/components listen attentively and unconditionally honor that position. From the place of hearing, understanding, and accepting all the parts, consensus or synergy can emerge. Where the collage approach brings a both/and tolerance for confrontational viewpoints to the table, the Way of Council is the mediation mechanism that helps manifest peace and harmony among them. It moves the potential for consensus out of the realm of wishful thinking and into reality.

This council process is nothing new. Ancient indigenous councils in all known civilizations have engaged such collective decision-making methodologies throughout recorded history and long before. Our ancestors have been sitting in tribal circles from time immemorial. Recently authors, change-agents, and visionaries have brought it forward again in works such as Jack Zimmerman's book *The Way of Council.*

In the council process, a group of people sit in a circle and pass a "talking piece" from person to person. The piece can be anything, even a simple stick. As each person receives the talking piece, he or she speaks as honestly and openly as he/she can, while the rest of the group listens from a place without judgment, comments, or questions. Heart-based listening and expression make the circle powerful. Leaving an empty seat at the table that represents the spirit of the relationship itself is another important aspect of the council process. When things are unusually difficult or confused, everyone can stop speaking for a while and listen to the spirit of the relationship. This respect for the relationship in question honors what the group is doing, even if people can't always hear what the relationship has to say—though usually, they do.

Many people have heard me say many times that we have to move from me-first to world-first on a global scale now to successfully heal our global climate crisis. How, logistically, can we do this? Council is a real,

These qualities (we can call them ego forms) are integral parts of our individual and collective wholeness, and we thus need to identify, honor, accept, balance, and include them in a greater unification toward healing our world comprehensively.

workable way for two or more entities: individuals, families, communities, businesses, states, nations, etc., or their constituents to *peacefully* communicate, through its listening and communication techniques, in order to reach comprehensive understanding and consensus.

Heart-based listening and expression make the circle powerful. Leaving an empty seat at the table that represents the spirit of the relationship itself is another important aspect of the council process. When things are unusually difficult or confused, everyone can stop speaking for a while and listen to the spirit of the relationship.

Consensus is akin to philosophic dialectic synthesis in that it unites opposing views holistically to arrive at a potential both/and solution. When groups larger than two people are involved, dialectic and dialogue can become multi-lectic and multi-logue. Listening to and respecting others is the key to reaching consensus, to creating a oneness of intention and purpose of many disparate viewpoints. Council is a tried and true way of accomplishing this.

Perhaps more than any other single human event or achievement, the NASA Apollo Moon program parallels what's asked of us all now. I have craved hearing from the Moon astronauts in their own words because, in many ways, the mission we are considering here is similar to our first landing on the Moon. It's similar in its audacity, its scale, its need for unanimous participation, and in the power of its ultimate effect. Both these efforts are things that have never been done before. Both involve

techno-scientific solutions *par excellence*. Both involve working together in unprecedented unity; both involve putting our mightiest hopes and dreams, our highest efforts and speculations to the test to see if we can do something successfully for the first time.

Perhaps more than any other single human event or achievement, the NASA Apollo Moon program parallels what's asked of us all now.

In this spirit of magnificent human accomplishment then, I hereby invite each and every one of us to call forth our deepest strength, faith, courage, and love for this greater unifying vision of our shared human, earthly, and celestial future as soon as possible—and to participate in it boldly, gleefully, authentically, with all the pride, humility, courage, strength, and honor we can muster.

THE COLLAGE APPROACH
AN INCLUSIVE, CONVERGENT PLANNING MINDSET

Our complex, pluralistic world has many varied and oftentimes competing interests in all the many fields of human activity and endeavor. This begs the question of how we can adequately and successfully address such complexity in making effective, world-healing plans. "Even the best-laid plans of mice and men" have often crashed and burned because two or more parties couldn't agree on how to proceed. To best include everyone and everyone's myriad needs at all the various levels we are about to consider, we would do well to foster an identification, understanding, honoring, and inclusion of all the parts (including all the potential adversaries).

Collage is a means, perhaps the only one accessible to us today, through which the spiritual and the

Neil Armstrong's first picture from
the Moon's surface of Buzz Aldrin
standing on the Sea of Tranquility.
We can see the U.S. flag and
the Lunar Excursion Module
reflected in his visor. What we
don't see here are the hundreds of
thoushands of dedicated workers
whose united collaboration
conspired to successfully put
these first two moon-walkers there
and bring them safely home again.
That's the sort of unified effort we
need to heal our world now.

empirical, the ideal and the real, the conservative and the radical, the dual, disintegrated, sacred/profane past and the unified, utopian ideal of a re-Eden-ized global future can be successfully woven together. As a design/compositional approach, it alone has the capacity for identifying, honoring, accepting, and including the essential nature of all relevant participants *on their own terms*, in order to best emblemize the intrinsic unity of all things.

In short, this inclusive, convergent strategy can incorporate the best of all worlds, mitigating all the many opposites in a unifying and inclusive, dialectic or multi-lectically synthesizing way. When we look at the many partisan worlds that co-exist today—in the realms of finance, science, religion, housing, govern-

In short, this inclusive, convergent strategy can incorporate the best of all worlds, mitigating all the many opposites in a unifying and inclusive, dialectic or multi-lectically synthesizing way.

ment, and politics, the military-industrial complex, environmentalism, the old and new energy economies, individual needs, collective mythology and participation mystique, etc., to name a few—it's easy to see how these many factors/sectors need to be successfully collaged.

When we consider through a collage approach the many varied yet convergent impulses to healing our world already in progress, we will be better enabled to pick and choose the solutions most relevant to any given circumstance. We will be empowered to marry any multitude of such healing models or modalities, with far less exclusive, me-first ideological angst. Perceived or imagined "incompatibility issues" will be less likely to hold us back from getting engaged and

making progress. We may even find that such an inclusive, unifying approach brings us into profound alignment with all of creation and our underlying contextual unity as well.

GENIUS LOCI

One final ingredient will complete the groundwork of specifying an archetypal prescription for healing our pluralistic, many-faceted world in realistic, do-able ways. There is another design sensibility which has long celebrated this truth. It is called *Genius Loci*, which means the spirit of place. Christian Norberg-Schultz, in his book by that name, spells it out in some detail. This approach to design and planning says, in brief, when in Rome, do as the Romans do.

I understand this approach to celebrate local forms and traditions in any sort of human endeavor. When we marry collage with genius loci and build from local, relevant form typologies, we can be here now while creating the best of all possible worlds for the future.

Voila! With the relationship diagram, the Way of Council, comprehensive planning through collage and genius loci methodologies, we have powerful change-process tools and a workable general formula for healing our world efficiently and tolerantly. I know from personal successes that each and every one of these tools can be an effective facilitator of positive change in the real, empirical world.

FIFTEEN

BRINGING IT CLOSER TO HOME

I spent my childhood summers in Ocean City, Maryland, where my best friend and I loved to ride the waves, as did the rest of my family. That was my first real experience of nature's awesome power, and the human ability to

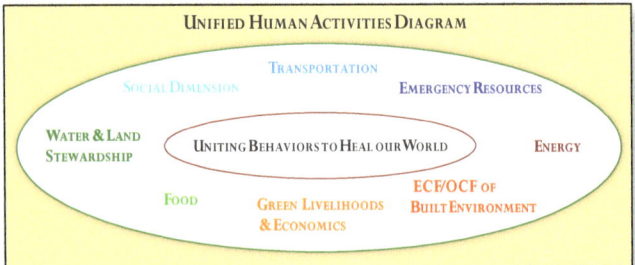

These areas of routine life impact Earth with carbon emissions and are where to focus our energy for beating the climate-change tipping point. This diagram helps us see these areas of our lives as a unified whole. We can focus on each one by turns in a comprehensive plan to lighten our carbon footprint. This approach (first developed during the design, construction, and habitation of our sustainable homestead) works at all levels from individual to global. Seeing our own presence here as a benefit of and benefactor to the larger whole of life on Earth, we naturally commit to healing that wholeness as best we can.

catch a "free ride." Back then, we dreamed of living the surfer lifestyle when we grew up, which meant doing the most with the least, leaving no footprints in nature (surfers of course ride waves and don't leave "footprints" in the water) and enjoying to the max our union with nature's rhythms. Everything I express about our transition to a low-carbon-footprint way of life shares that same spirit, and I'm still just as stoked about living a low-impact life now as I was back then.

When I give sustainability talks, people sometimes come up and say something like: "That's great, Mr. Temple. I'd love to do all the things you mentioned, but, you know, I have a family to support, bills to pay, and well, I just can't afford either the money or the time it takes to do what you've done. How did you ever manage it?"

This obvious and significant practical issue is partially addressed today by new opportunities to integrate green livelihoods and our lives. Green jobs are multiplying ex-

ponentially and many do not require years and years of training. They include performing energy audits, installing solar equipment or wind turbines, designing and building sustainable homes and commercial buildings, rehabbing homes to increase their energy efficiency, green architecture, green product design (affecting countless areas of manufacturing), resource management (locally, nationally, or internationally), sustainable agriculture (locally or internationally), pollution cleanup (including research into the best ways to accomplish it), etc.!

When people lose their jobs and homes, as many have in the recent Great Recession, it feels as if the earth has slipped out from under them, that life is built on sand rather than bedrock. I found that building a different relationship with my homestead helped me find bedrock. Carol and I are the kind of people who connect deeply to land and home, and our bond with the land on which we live is a potent source of inspiration and empowerment. Home ownership strengthened our sense of self-reliance and all that comes with it. Owning our own source of external power enhanced our ownership of our inner or personal power. We took increasing responsibility for all the details of our lives. We owned the means of growing our food, which gave us control over what goes into our mouths, our health, our bodies, and the way we interact with the land, water, and air. We have benefited hugely from this.

The homesteading experience is the central unifying element in my role as an Earth-healer. Spanning the past two decades, it was the cradle of many of the seminal ideas presented in my Global Awakening series, and the first place where I put all my theories about unity to the test. There's much more to say about lessons learned here, the elements and principles involved in living the low carbon life, etc. Yet, to keep with our "Glimpses" theme, I'll save the details for the appropriate *Global Awakening* series, *Part III, Next Steps* volume.

SIXTEEN

THE WAY OF QUIET

I can't resist giving you just a couple more glimpses though. Crestone is a small town on the edge of the wilderness and is extremely peaceful, as you might imagine. The quiet we experience here is one of the key elements of our *genius loci*, the spirit of our place (together with its important companions: spaciousness, nature's visual beauty, and the clarity and openness of the atmosphere here). The quiet plays many roles in my life, and I believe it can create a positive feedback loop that leads to all sorts of healing. That's why the ancients had their folks go out into the desert for visioning and healing retreats. Among the precious attributes that such questing could engender, natural quiet can bring transformational change and rare gifts such as the greater and more lasting inner experience of peace of mind.

The peacefulness and quiet of a natural setting such as this can soothe the soul.

Eventually, the external quiet can work deep within us, providing yet deeper gifts. Once we peel away layers of our life's inner challenge(s), we can come to a profound sense of integration and wholeness. Living more quietly can also translate into greater environmental care, and a presencing of peace and calm in the face of external anxiety and turmoil too. The Way of Quiet can be a natural way of healing for self, others, and the world.

Of course, you don't have to live in a remote village to have silence in your life. It is one of the many gifts that comes from adopting a transformative, Earth-healing way of living, wherever you are.

One of my very favorite places on this Earth! When I go out walking, I usually end up sitting by this stream, in these woods. Its soothing sounds make it a great place to think, observe, listen, contemplate, and pray.

Lovely blossoms of early spring

The beautiful colors of our landscape in the fall

What I love the most about our landscape in the fall is the wonderful smells that happen through the special unity created among the dead and the living beings, the sunlight, and the moisture. Many plants turn beautiful colors and frame nice views of the landscape, too.

Summer beauty in our flower garden

Winter is a time of rest, inspiration, peace, and tranquility, of sitting in contemplation of nature's power, beauty, and presence.

SEVENTEEN

GIVE SOMETHING MORE THAN YOU TAKE

This motto, which I first learned at McDonogh School in Baltimore in the '60s and '70s, stayed with me and translated into sharing Earth's bounty by giving gifts to family and friends. Carol received similar guidance from her family when she was growing up on the west coast. In my life now, I continue the tradition and respectfully follow our role models, the mighty cyanobacteria. My family now gives something more back to the Earth, our friends, neighbors, and the world than we receive. We give our carbon-neutral electrical power generation back; we give a better atmosphere back; we give food back; we give a peaceful, calming presence and beautiful gardens back. We encourage more flora and fauna couplings and pairings, more offspring and a flourishing plant/animal habitat, all of which enriches our lives, our world, and life all around us.

We give our carbon-neutral electrical power generation back; we give a better atmosphere back; we give food back; we give a peaceful, calming presence and beautiful gardens back. We encourage more flora and fauna couplings and pairings, more offspring and a flourishing plant/ animal habitat, all of which enriches our lives, our world, and life all around us.

The Tea Garden in San Francisco's Golden Gate Park.

Carol and I planted and watered the honey locust seeds we received from the Dalai Lama at the end of the 1991 Kalachakra initiation I mentioned in Chapter Eight. They grew, and we picked the strongest sprout, which grew more. When we moved to Colorado in 1993, it just barely fit into the car. In 1994, we planted it in our new garden, where it continues to grow. And grow! Today it is a huge part of our beloved nature sanctuary. I put an Adirondack chair under it several years ago, and I sit there on sunny summer days, enjoying the beautiful garden we have created and the shade afforded by HHDL's gift to us. That tree, like much of the natural world, gives us far more than it takes.

EIGHTEEN

UNITY IN THE COMMUNITY

After moving to Colorado from Ithaca, it wasn't long before the community organizer in me began to say hello. In the 1990s, Crestone used to put on what we called the Cabin Fever Talent Show each February, a welcome relief from being cooped up all winter due to deep snow and freezing temperatures. About twenty years ago, I worked up a skit for that year's event starring an invented character, the Mystical Cowboy, and focused on the mantric theme of "unity in the community."

I developed this skit in part for fun, but also because our town was horribly fractured and polarized at that time. People had actually been reduced to throwing chairs at each other at town meetings and other events, not a great picture of American democracy at work. Miraculously, my skit shined some levity on our plight (a recurrent theme in the history of democratic government). This was the first time I felt called to speak on unity so clearly and forcefully. From that point on, it powerfully gripped my deepest consciousness, sometimes rising to the foreground of my life and sometimes receding into the background.

Later, I used that theme in community organizing, gathering various stakeholders and interests (food, energy, etc.) together, but I'll save that story for another volume. The sketch version of it is that building unified, sustainable community takes many important ingredi-

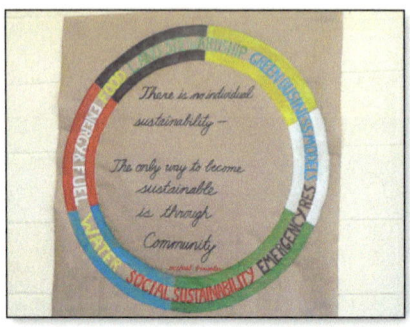

Crestone's Sustainability Conference theme poster, drawn by Jo Anne Kiser. We used the same unity of parts approach to advance community sustainability.

Grass roots efforts such as ours, directed and facilitated by local environmental champions and gathering broad community participation, will do much to bring along governments that are either averse to such changes or unable to enact them legislatively.

ents. One of the most important is the ability to find and integrate committed stakeholders and dedicated champions—people who keep their eye on the ball every day, week, month, and year, and don't back down or give up, to accomplish community goals. Further, it takes getting these committed champions into a room where they sit down together to map out a comprehensive strategic plan for the future that integrates all their efforts into a holistic approach.

If you decide to unify your community around the theme of healing our world, finding and nurturing those who are natural champions of Earth healing will make a big difference in your community's successful progress towards its goals.

Grass roots efforts such as ours, directed and facilitated by local environmental champions and gathering broad community participation, will do much to bring along governments that are either averse to such changes or unable to enact them legislatively. This unifying, comprehensive, consensus-based approach is workable for any scale of community. Such efforts and their outcomes help communities see that they can overcome the nay-saying me-first resistance and all the other enormous challenges to change, and instead manifest their vision for this greatest time in human history, a time without precedent, when so many are joining together to transform our world.

NINETEEN

In preparing for a sustainability conference I coordinated in Crestone several years ago, I approached the Crestone Charter School (CCS) to see if teachers and students would like to help discover Crestone's real carbon footprint. It became a school-wide special project that began with a meeting outlining the evolution of our current climate problem, as well as a discussion of both the rights and liberties we enjoy and the real and grave responsibilities for stewarding our precious gifts on behalf of this and future generations. At one point, I looked around and noticed that the room was totally silent and we had everyone's complete attention. These kids of many different ages really got it, and they jumped in enthusiastically, wanting to help any way they could.

Over the next month, under the direction of my good friends Vince and Mary Palermo, the students gathered data on things like how many greenhouses, grow domes, chickens, and eggs we had, where we got our food, and how much coal we were responsible for our local utility burning annually. Though Crestone has long been branded as a spiritual and sustainability Mecca, our students discovered that our community of roughly 1100 souls emitted (as of June 2007) 37 million pounds of carbon per year, or 91 lbs./person/day. We used over half a mile's worth of railroad coal cars filled to overflowing to generate our power every year, and a mile's worth of propane trucks for our yearly heating and cooking. The students found that we consumed a staggering 1.5 million pounds of food per year, most of which was trucked into town in around 100 semi's. Some of it came from as far away as New Zealand. We put about the same

The gathering for the grand opening of the Crestone Charter School, a cutting-edge example of carbon-neutral school design, with Challenger Peak in the background.

weight of garbage into the landfill each year, too. What a wake-up call!

Crestone's children taught us adults a great lesson that summer. They showed us the sobering truth (and valuable grade-school lesson) that our actions still spoke louder than all our green rhetoric ever could.

Crestone's children taught us adults a great lesson that summer. They showed us the sobering truth (and valuable grade-school lesson) that our actions still spoke louder than all our green rhetoric ever could. Then this initial mutually respectful and supportive contact with CCS students unexpectedly became the seed of something much bigger. The CCS faculty and board became more strongly aware than ever of the value of environmental consciousness/sustainability. This supported their choice, several years later, to opt for carbon neutrality in the master plan for their new school campus.

On September 4, 2012, CCS founder Karen Acker (pictured here on the left, in the school's new solar courtyard) cut the ribbon to officially open the new CCS building. Joining her were members of the CCS class of 2013, the first students who graduated in the new structure.

Much to everyone's astonishment, CCS applied for and then *actually landed* a $6+ million BEST ("Building Excellent Schools Today") grant to build the new school. With other key community stakeholders, I was invited to participate in the new design process, keeping those carbon-neutral goals on the front burner all along the way. The CCS board chose not only to design a 100% carbon neutral campus (and the built facility is quite close to 100% for its power and heating needs), but also to facilitate a closer connection between students and their natural environment on a daily basis. The kids who started attending CCS on its new campus in September of 2012 will grow up living the new paradigm exemplified by a school that, for them, was always an educational environmental leader. Hallelujah!

The broader message here is that our students and teachers are starving for opportunities to make a palpable difference in their lives and the state of our world. They profoundly need to see us living in integrity with what we teach. They want to see us taking actions that match our words, as we try to shape their understanding of their relationship to the world they will inherit and steward for their children and all sentient beings.

In our case, actions have truly spoken louder than words. The ribbon-cutting opening ceremony for the CCS occurred on September 4, 2012. I was in attendance together with many other proud and enthusiastic community members, and it was truly an inspiring and heart-warming day for those of us who have long harbored such dreams. The school turned out incredibly well, thanks essentially to a process that included everyone at the planning table, all the numerous, deeply intertwined contributory eforts by the many good folks involved, namely: The Colorado Department of Education, Colorado Attorney General's Office, the League of Charter Schools, the Moffat School District, the Town of Crestone, outstanding planning and design by Harry Teague Architects and associates, exceptional design/construction coordination/leadership by Marie-Louise Baker, the CCS Governing Council and Design Committee, DSI Construction, many skilled subcontractors including several extraordinarily talented local alternative energy subcontractors, literally hundreds of other local community contributors, and of course, the CCS administration, faculty, staff, parents, and students themselves.

Our great team effort has guaranteed an excellent design and built environment for our children for many years to come. Today CCS is over 95% of the way toward full neutrality. Its 24kw pv solar array, together with the innovative deep solar thermal heating system, will eliminate at least 36 tons of carbon yearly, or 16 tons of coal. The future installation of a solar-powered methane digester to replace the installed back-up propane system will eventually make it 100% carbon-neutral, whenever such a change-out happens in the future. This was truly a great and unprecedented moment for our community. It provides compelling evidence that, with a powerful vision and inclusive and holistic teamwork across many levels and stratums of society, such meaningful, comprehensive planning can and does happen, and can yield exceptionally wonderful fruit.

TWENTY

UNITING FOR THE COMMON GOOD

Whatever we put our attention on, grows. Just imagine the powerful effect that regular daily or weekly prayers for the Earth's well-being could have on our sense of community and experience of unity with all humanity and every aspect of the universe. This is a tremendous opportunity for our thoughts, feelings, and actions to reflect the idea that all parts of this marvelous earthly realm are God's essence.

We can find a personally relevant and authentic way to merge our spiritual yearnings for union with God (or whatever name we call our divine Source) with the recognition that Earth and all her inhabitants are expressions or reflections of one Source, and support their well-being as we do our own and that of our families. We can commit to engaging in this as soon and as often as possible, and with as much love and presence and oneness-awareness as we can muster. The future depends upon everyone seizing this moment and discovering how truly powerful we can be as one voice.

Our precious and fragile human world is always renewing itself, as is the greater universe. Our contemporary human existence is the product of billions of years of myriad steps of evolutionary transformation. Both the timeless One and the temporal Many are meshed and collaged together to form the basics of our living reality, and now we

This is a tremendous opportunity for our thoughts, feelings, and actions to reflect the idea that all parts of this marvelous earthly realm are God's essence.

know how to work with this truth in all our efforts moving forward to heal our world.

We're now broadly converging in our different approaches to healing our self-created global malaise. This is so, even as we've still much to do to move future generations and our Earth's climate into the safe zone again. The holistic, integrative approach to personal and global healing championed in the Global Awakening series does work at all levels of human life. It can be employed in our moment to moment awareness, as we live and work in our daily lives. By asking what we want our lives and world to feel like, by honoring all the parts and wholes and their interrelationships, and by making comprehensive plans from these key ingredients, great healing can emerge. This especially includes the unification of becoming and being that allows us to align our behavior more fully with the soul of creation. Its authentic creative mechanics are the true ground of shared wholeness underlying every aspect and form of the universe's evolution.

TWENTY-ONE
THE GRAND UNIVERSAL LINEAGE

Because one broadly accessible, workable way of reasserting our wholeness embraces the many contributory elements that make our fragile human existence possible, I find it quite important to deeply investigate the nature of the universal creation process. Such inter-relationships have traditionally been known in various cultures, fields, and time periods as *lineages*.

In scientific terms, a lineage is a subset of the evolutionary tree of life. Each is generally thought of as a sequence of species that form a line of descent, with each new species emerging from its immediate ancestral species.

Various scientists, particularly Jared Diamond, see human wisdom-transmission from elder to younger as a key evolutionary ingredient of our species' survival, progress, and transformation. The reverence for ancestors we see in indigenous and other philosophical, spiritual, and/or religious traditions is likely a recognition of the importance of this capacity in us to pass on wisdom.

Indigenous spiritual leaders in traditional cultures the world over—notably the lamas of Tibet and the Vedic pundits of India—can recite the names of their predecessors, and often do so before public speaking or teaching. They do this not only to honor them but also to acknowledge their forebears' living presence in their own lives. These unbroken successions of wisdom-keepers often extend back hundreds or even thousands of years. The same is true for some royal family lineages. Such lineage reverence echoes the religious admonition to "Honor thy father and mother."

One broadly accessible, workable way of reasserting our wholeness embraces the many contributory elements that make our fragile human existence possible.

Traditional wisdom teachings never had the worldwide exposure that is now possible, nor could most have had the trans-cultural acceptance that a scientific picture of reality enjoys today. Contemporary science makes it possible to clarify and understand our ultimate common unified lineage—the one that takes us back beyond the birth of humanity, the Earth, the stars and galaxies, all the way back to the dawn of time. This is a truth long held by many spiritual traditions that could not prove or present it in any universally acceptable terms.

Thanks to startling recent developments in inter-related scientific fields from astrophysics to earthly biology, genetics, and anthropology, we can now more effectively trace the lineages of all beings, places, and things. Whether it's atoms or arachnids, galaxies or gold, regular folks like you and me or emperors, high lamas, gurus, and indigenous elders, all forms of manifestation are clearly linked to a single shared source in Originating Power and its Big Bang universal origination event. Known by many names the world over, Originating Power is the ultimate great-ancestor of all wisdom traditions, racial or ethnic groups, and even humanity itself.

I call this complete, unbroken chain of events and phenomena that has enabled the existence of ourselves and our shared world our *Grand Universal Lineage*. It includes, honors, and unites all beings, places, and things, everywhere.

The Grand Family Tree of Existence drawing by Emily S. Damstra. All major precursor aspects of this "Grand Universal Lineage" have enabled our earthly existence today. It thus shows our underlying Contextual Unity with the rest of the universe.

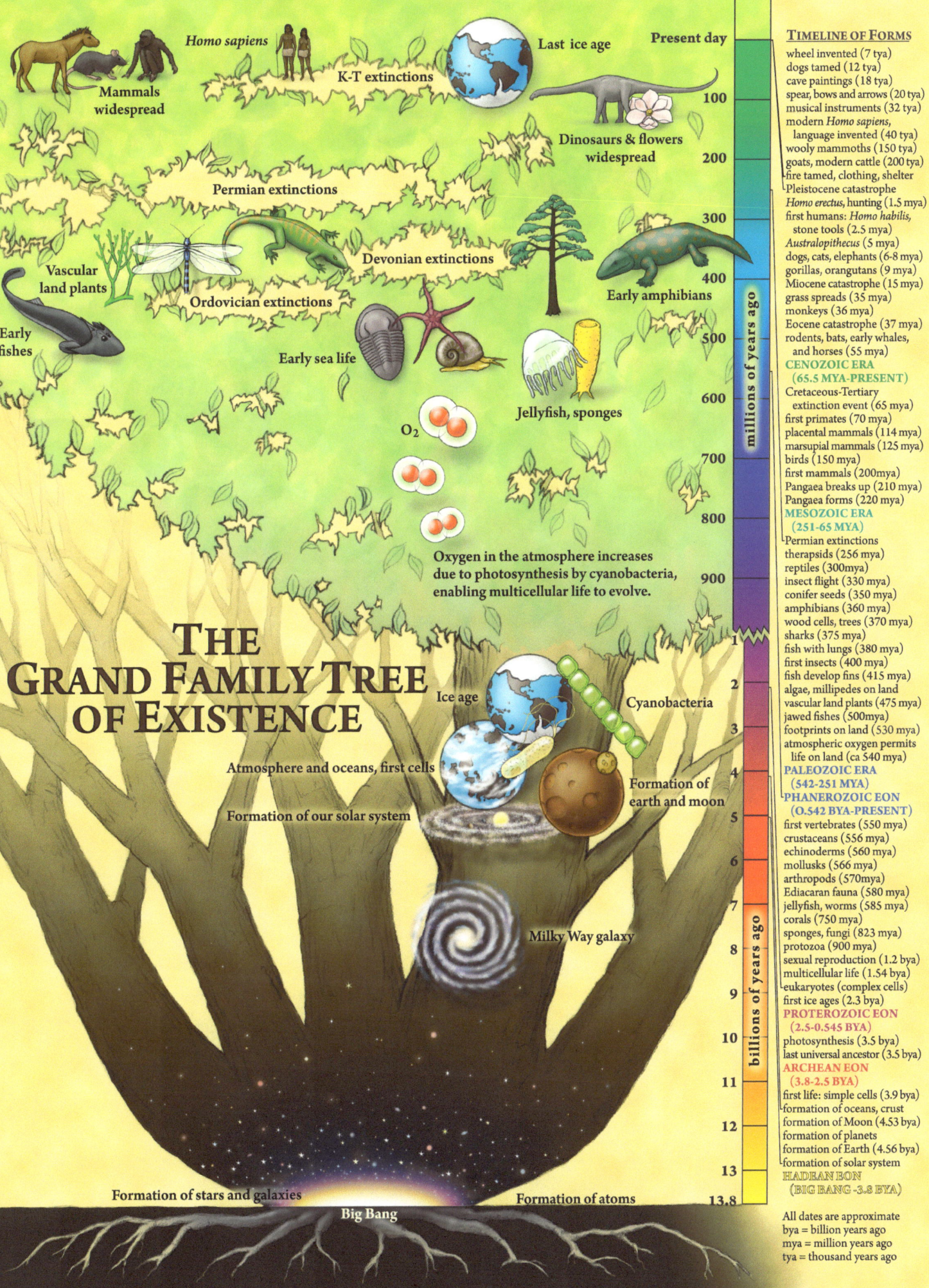

THE GRAND FAMILY TREE OF EXISTENCE

Mammals widespread

Homo sapiens

K-T extinctions

Last ice age

Present day

100

Dinosaurs & flowers widespread

200

Permian extinctions

300

Vascular land plants

Devonian extinctions

Early amphibians

400

Ordovician extinctions

Early fishes

Early sea life

500

Jellyfish, sponges

600

O_2

700

800

Oxygen in the atmosphere increases due to photosynthesis by cyanobacteria, enabling multicellular life to evolve.

900

Ice age

Cyanobacteria

Atmosphere and oceans, first cells

Formation of earth and moon

Formation of our solar system

Milky Way galaxy

Formation of stars and galaxies

Formation of atoms

Big Bang

millions of years ago

billions of years ago

1

2

3

4

5

6

7

8

9

10

11

12

13

13.8

TIMELINE OF FORMS

wheel invented (7 tya)
dogs tamed (12 tya)
cave paintings (18 tya)
spear, bows and arrows (20 tya)
musical instruments (32 tya)
modern *Homo sapiens*, language invented (40 tya)
wooly mammoths (150 tya)
goats, modern cattle (200 tya)
fire tamed, clothing, shelter
Pleistocene catastrophe
Homo erectus, hunting (1.5 mya)
first humans: *Homo habilis*, stone tools (2.5 mya)
Australopithecus (5 mya)
dogs, cats, elephants (6-8 mya)
gorillas, orangutans (9 mya)
Miocene catastrophe (15 mya)
grass spreads (35 mya)
monkeys (36 mya)
Eocene catastrophe (37 mya)
rodents, bats, early whales, and horses (55 mya)
**CENOZOIC ERA
(65.5 MYA-PRESENT)**
Cretaceous-Tertiary extinction event (65 mya)
first primates (70 mya)
placental mammals (114 mya)
marsupial mammals (125 mya)
birds (150 mya)
first mammals (200mya)
Pangaea breaks up (210 mya)
Pangaea forms (220 mya)
**MESOZOIC ERA
(251-65 MYA)**
Permian extinctions
therapsids (256 mya)
reptiles (300mya)
insect flight (330 mya)
conifer seeds (350 mya)
amphibians (360 mya)
wood cells, trees (370 mya)
sharks (375 mya)
fish with lungs (380 mya)
first insects (400 mya)
fish develop fins (415 mya)
algae, millipedes on land
vascular land plants (475 mya)
jawed fishes (500mya)
footprints on land (530 mya)
atmospheric oxygen permits life on land (ca 540 mya)
**PALEOZOIC ERA
(542-251 MYA)**
**PHANEROZOIC EON
(0.542 BYA-PRESENT)**
first vertebrates (550 mya)
crustaceans (556 mya)
echinoderms (560 mya)
mollusks (566 mya)
arthropods (570mya)
Ediacaran fauna (580 mya)
jellyfish, worms (585 mya)
corals (750 mya)
sponges, fungi (823 mya)
protozoa (900 mya)
sexual reproduction (1.2 bya)
multicellular life (1.54 bya)
eukaryotes (complex cells)
first ice ages (2.3 bya)
**PROTEROZOIC EON
(2.5-0.545 BYA)**
photosynthesis (3.5 bya)
last universal ancestor (3.5 bya)
**ARCHEAN EON
(3.8-2.5 BYA)**
first life: simple cells (3.9 bya)
formation of oceans, crust
formation of Moon (4.53 bya)
formation of planets
formation of Earth (4.56 mya)
formation of solar system
**HADEAN EON
(BIG BANG -3.8 BYA)**

All dates are approximate
bya = billion years ago
mya = million years ago
tya = thousand years ago

We can better understand the value and importance of the Grand Universal Lineage thusly: We each individually (you and I, for example) could not exist as we do without the prior existence of our own personal family ancestries, all of which couldn't exist without the prior existence of humanity, which couldn't exist as it does without the prior existence of earthly life, which couldn't exist without the prior existence of Earth, which couldn't exist without gravity and all the large celestial forms (like the Sun) that it birthed [midwived?], all of which couldn't exist without the prior existence of the other universal laws and forces of nature that apply to elementary particles, atoms, and so on.

If we deeply consider all these interrelated contributory parts and the living, vibrant whole they compose, it's hard to escape the conclusion that each and every aspect in the Grand Universal Lineage is a major and pivotal achievement of universal evolution—and that they and our vast cosmos are indeed miraculous, sacred occurrences.

Science has revealed and verified the specific mechanics of this amazing universal form continuum to an unprecedented extent in this and the previous few decades. The invention of the Internet in 1989, made possible by previous innovations such as the computer, TV, telephone, and telegraph, has also provided us the unprecedented capability of sharing knowledge of our singular lineage with everyone in the world.

Contemporary scientific verification of the ancient spiritual vision of a single common lineage begs a closer look by the world's scientific, indigenous, religio/spiritual, and philosophic traditions, so that we

From our Grand Universal Lineage, we gain the crucial awareness of our ultimate kinship. As human beings, we are all related—as inter-related parts of an immense, living, evolving whole whose nature and mechanics we can all now hold much more clearly in our understanding than ever before.

might all find in it effective ways to establish a deeper common ground for our teachings and Earth-healing efforts moving forward.

From our Grand Universal Lineage, we gain the crucial awareness of our ultimate kinship. As human beings, we are *all* related—as interrelated parts of an immense, living, evolving whole whose nature and mechanics we can all now hold much more clearly in our understanding than ever before.

We should care about deepening this holistic sort of existential understanding, because how we conduct ourselves in the immediate future will affect the long-term viability not only of our species but of Earth's entire biosphere as well. Just as each successive development in the Grand Universal Lineage is made possible by its ancestors, so too our future and the future of earthly life depend upon us.

This lineage's ultimate defining quality of immense interrelatedness can also help us better revere all life. It helps us see. for example, that we humans are not possible without our world-based support system. Further, we see that our behavior has an equally profound effect on our "Mother" Earth and her biosphere today. Our individual actions can be seen as contributory parts of the larger whole of our collective human behavior—as essential dynamics in the meaningful mosaic that is human nature, itself a small and now-critical constituent of Mother Nature.

Embracing the Grand Universal Lineage's intrinsic interdependence and non-autonomy can also help us solve global problems such as climate change. Inter-relational awareness can help us accept and better integrate the many contemporary Earth-healing efforts (such as large- and small-scale green energy systems, top-down legislation/regulation and bottom-up grass-roots efforts, etc.) more holistically and harmoniously, which will improve their effectiveness moving forward.

The Grand Universal Lineage reveals our ultimate shared ancestry and provides contemporary scientific verification of the ancient spiritual truths that all life is sacred and that our fragile human existence relies on something much bigger and more powerful than even our strongest governmental, corporate, or military machinery.

Understanding our Grand Universal Lineage helps us see that many different scales of activity can mesh together harmoniously in authentic, mutually respectful partnership. Internalizing the bigger picture of our inter-relatedness (and common vulnerability) could facilitate a lasting transition past the petty politics of our partisan persuasions into constructive collective cooperation and decisive, Earth-healing action. This is valuable in our urgent time-frame as we need to quickly find and implement all workable solutions.

The Grand Universal Lineage reveals our ultimate shared ancestry and provides contemporary scientific verification of the ancient spiritual truths that all life is sacred and that our fragile human existence relies on something much bigger and more powerful than even our strongest governmental, corporate, or military machinery. (We wouldn't be here without the rest of creation, and whatever happens to life on Earth happens equally to us.) It also gives us a larger holistic conceptual framework that, when applied

to our own behavior, can help us better engender decisive change and celebrate it in the Earth-healing contributions of others.

This extended lineage sensibility also shows us our specific place in the bigger picture: As our ancestors gave us life's treasure galore, our great-grandchildren's fate is tied to every step we take; to our daily interactions with family, community, and world. We are *their* ancestors. Will we abandon or save them?

TWENTY-TWO

SEEING LIFE'S BIG PICTURE

The part of our Grand Universal Lineage that we all love most is undoubtedly our own species. And one of humanity's greatest qualities is our proclivity for perpetual change. Our species can boast some real movers and shakers! Over our known (and conjectured) history, we clearly weren't content just lounging around eating bananas. We stood up, walked and rode, tamed and hunted, farmed and fished, danced and flew, built and destroyed. We invented, communicated, made love, war, and peace; we went to the Moon and photographed the star-filled universe. We produced and presided over the most rapid and comprehensive transformation in life's known history. We have done so well in so many ways, but at the same time, we've also put our lives and those of all other earthly sentient beings at risk.

As I researched the story of our planet and species in developing the *Global Awakening* series, I received a striking revelation. I saw the evolution of life *as a whole*. I saw it as a unified phenomenon; the manifestation of a single living organism and the completeness of life containing it, bound together inexorably as one and the same being. Evolution's individual stages, as we've experienced them here on Earth—amoebas, flowers, trees, mammals, dinosaurs, us, etc.—appeared to me as individual frames in a moving picture, functioning as both discrete living elements and integral parts of a larger, organic whole. It became abundantly clear that humanity, along with every other expression of life, is just one of the innumerable elements of this wholeness. Every element is a unique energetic presence and form, and all simultaneously express the underlying presence of Innate Life Intelligence.

Just as the individual shots in a movie come together and contribute to a bigger story, all the forms that have appeared in our planetary history tell a tale that is bigger than the sum of their individual contributions. Together, they narrate the grand journey of cosmic presence into palpable, living form: the story of life on Earth.

> *Just as the individual shots in a movie come together and contribute to a bigger story, all the forms that have appeared in our planetary history tell a tale that is bigger than the sum of their individual contributions. Together, they narrate the grand journey of cosmic presence into palpable, living form: the story of life on Earth.*

Right: The Grand Family Tree of Humanity, illustration by Emily S. Damstra.

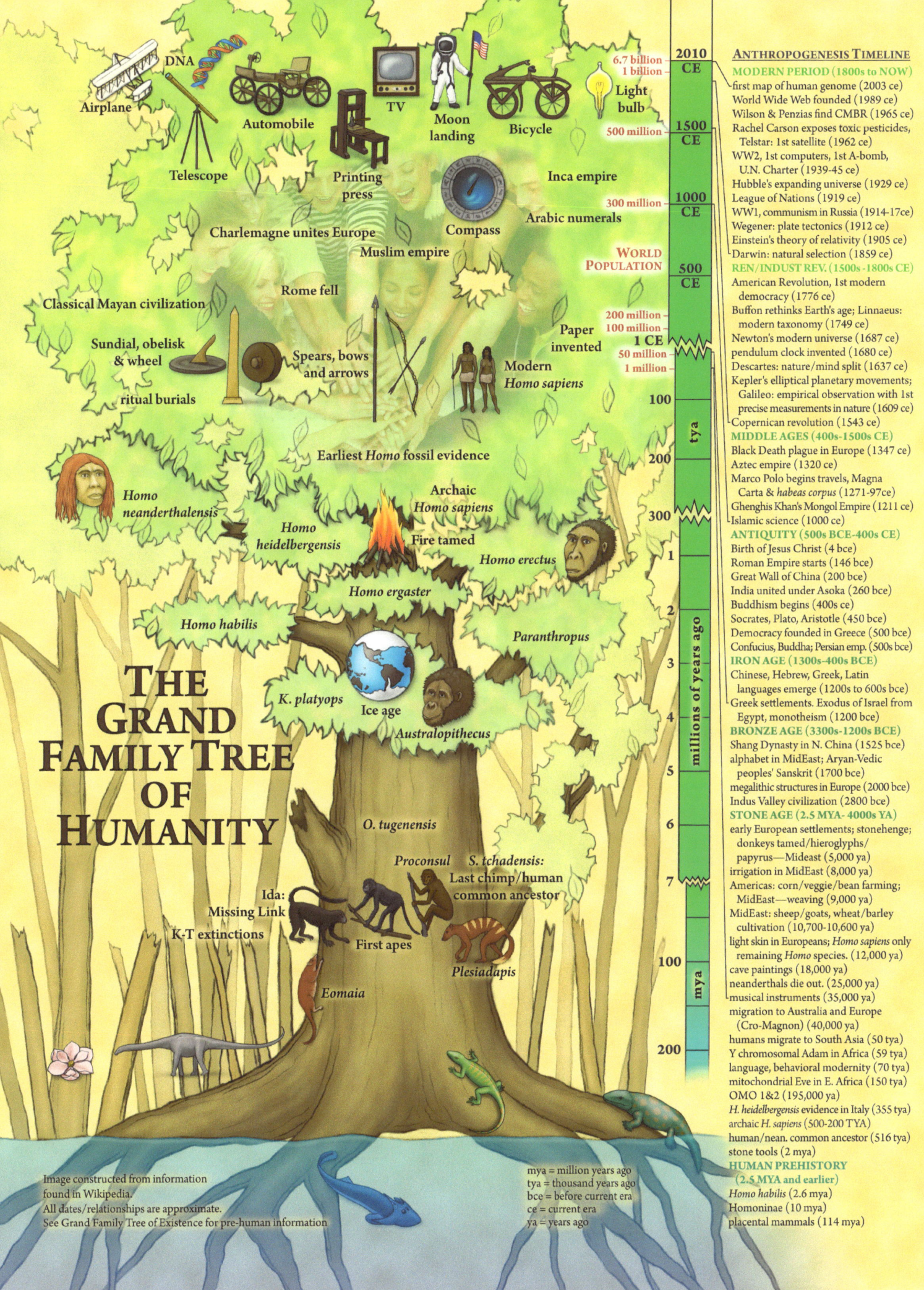

TWENTY-THREE
MAKING IT PERSONAL

My research has revealed a way that we can all relate to the essential interdependence of our part in Earth's life-story directly and personally. Hold up your primary (writing) hand in front of you and focus your attention on it. Turn it back and forth, then flex it up and down from the wrist. Study it as you would a work of art that you see for the first time. Notice the lines, the skin's delicate patterns and imprints, the joint wrinkles, the color and sheen, the veins visible just under the surface. With your other hand, squeeze and feel the bones beneath the skin of the first. Flex and fold your hand and feel yourself present in it; feel the life and power it embodies.

Remember that your hand, like every part of your physiology, is constructed with atoms that once were part of other human beings during every stage of human development in the past seven million years. That's right. You're actually looking at and holding our species' entire seven-million-year saga right in the palm of your hand.

Our bodies literally are *Proconsul*, Neanderthal, *Homo erectus, Homo habilis*, and all the rest of the incredible cast of characters that has populated Earth. The atoms in our bodies have dwelled in every human race, on every continent, and in the citizens of every civilization. Our hands contain atoms from our most revered saints and scientists and our most notorious scoundrels, from the wealthy and the poor, the farmers and the city folk, and ancient friends and enemies. These same atoms that have journeyed through so many human forms have also been part of every other

> *Remember that your hand, like every part of your physiology, is constructed with atoms that once were part of other human beings during every stage of human development in the past seven million years.*

structure in existence from the beginning of time, all thanks to the mighty explosion of cosmic possibilities that began with the Big Bang.

I invite us all to remember this authentic aspect of our presence in the here and now, and the fact that we owe our very existence as a species to this immense, single living web of evolutionary coexistence. That means that we also owe our existence to many other beings and forms, processes and products of previous evolutionary development that coexist with us. And we've seen that accurately appraising the scope and nature of such an existential web/debt can form the basis of our contemporary global awakening and transformation toward more convergent and compassionate world-first healing behaviors.

How does this actually play out in our individual lives? In my case, this gift of life and its associated debt involved the utmost sacrifice by an ancestral family member, my father's brother. He is my namesake, and although I've never met him, in some important ways, I owe and have dedicated my life to him. He grew up with my Dad in the nineteen-thirties, was a talented trumpet player, and when our nation's "rendezvous with destiny" arrived, they both went off to serve our country in WWII. Uncle Lee trained for and became a fighter pilot, and he led his squadron of P-38s over France to provide air cover for the fateful landings on the beaches there on D-Day, June 6, 1944. He gave his life leading his men in this important effort, and received the Purple Heart for his courageous valor that day.

I was born twelve years after he died, and my parents named me after him. On a deeper level, his brave sacrifice made it possible for me to exist, and ultimately, to write this book (living a precious life that was taken from him). I often remember Uncle Lee and his supreme courage when I have to face my life's greatest challenges. No matter how hard or big they are, they're peanuts compared to what he and so many others did for me, for all of us. All these sorts of beings and events in all our families are linked, just as our own contemporary sacrifice for future generations and all of life through our heroic dealing with the climate crisis today is also linked to their being spared a life of unspeakable misery and devastation tomorrow. In other words, we've learned that recognizing and honoring such profound linkages is a vital aspect of what living in unity is all about.

There are many other aspects of my life, my own evolutionary story, that made this work possible, and thus, they are all a part of my intrinsic unity with you and with all that is. Our focus on oneness does not take away from the importance of the many varied parts, it's just important to keep sight of the whole *in addition* to seeing the individuality of each of them, simultaneously—holding the relevance and value of both parts and whole, Many *and* One, *together*.

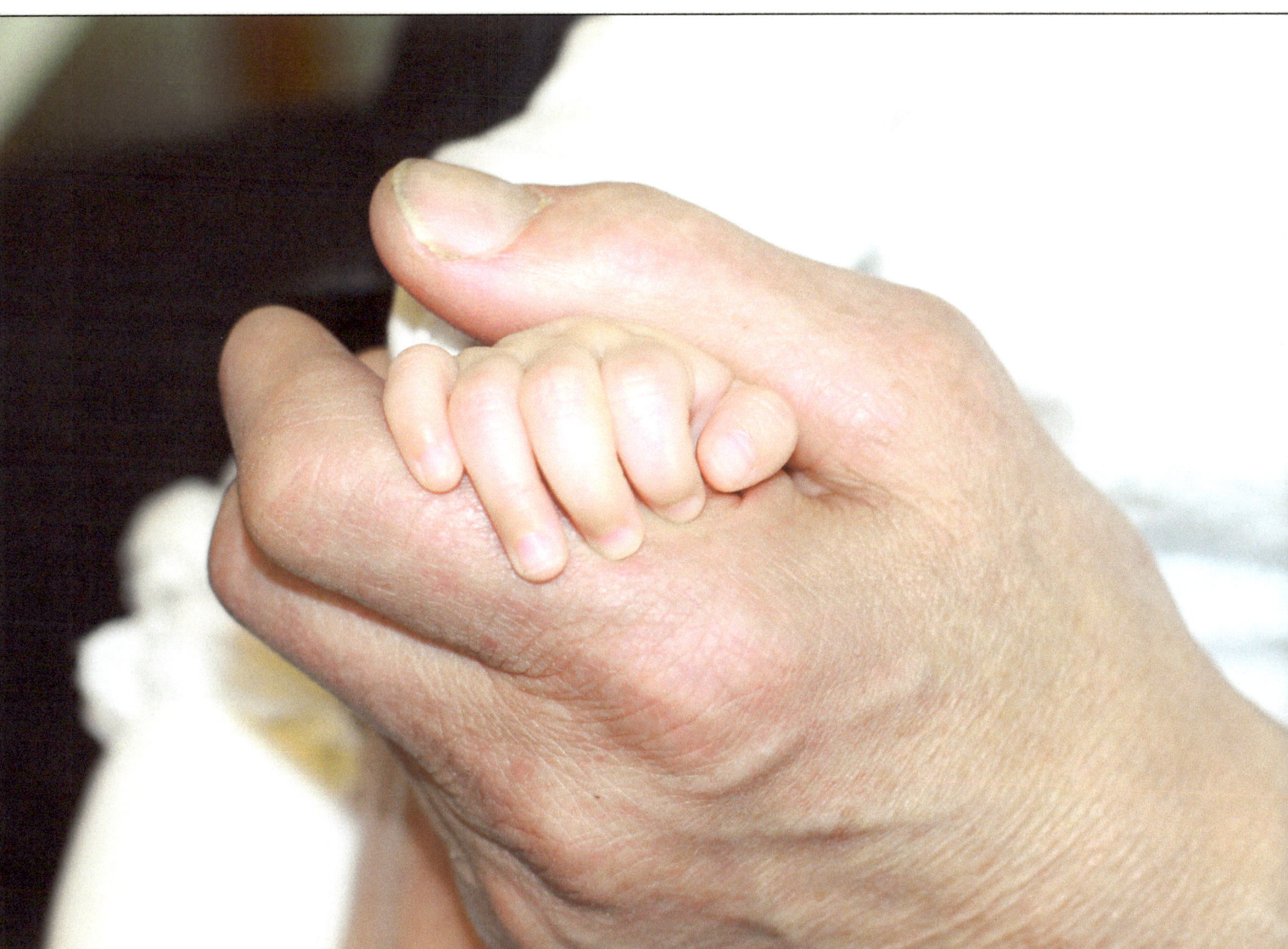

EPILOGUE

I'VE HAD A DREAM!

Joseph Campbell once said that dreams are personal myths and myths are collective dreams.[8] The following example brings this powerful truth and its interweaving of personal experience and planetary mythology to life. Over a decade ago, a few years before the seed of this massive body of work unshakably planted itself in me, I had a really memorable dream. It was so powerful that every time I think of it or mention it to anyone, I get chills and goosebumps!

In the dream I was standing on a stage in front of a crowd of perhaps a couple thousand people. There were some other folks standing on the stage with me, but as I glanced around, I didn't recognize anyone. I had just been introduced to the audience, and I was standing behind the lectern and its microphone, so I was obviously about to say something. Yet I had absolutely no idea why I was there, or what I was about to say! The audience sat looking at me in silent, rapt attention. As you might imagine, I was terrified! I'm usually okay with talking to groups, but only if I have some basic sense of what's to be said. This time I didn't have a clue, and they were all sitting there quietly, waiting for me to speak. (If this happened in real life, now is when I'd start to pee my pants!)

As I pondered this bizarre situation, I felt a tap on my shoulder. I turned around to see Reverend Martin Luther King, Jr. standing right next to me. He smiled at me in friendship, and said: "Lee, I'm really looking forward to hearing what you have to say, because it's really important, and you're the only one who can say it." He smiled lovingly at me again, gave me a little pat on the back and, nodding his head toward the audience (as if to say "they're waiting for you" or "go for it!"), he stepped behind me with the others on the stage, so that I could start speaking. That's when I woke up (in more ways than one)!

Now, for most of us, Reverend King is the very pinnacle of great and powerful, truth-speaking oratory in action. So for him to say this to me, even (make that especially) in such a personally oriented dream, has been a life-changing event for me.

Suffice it to say that now I have a little better idea of what I'm saying and to whom! But also, this dream surely helped empower me in many ways to speak the truths I share throughout the Global Awakening series. Remembering it from time to time has also helped keep me going on to the finish, even after I discovered that the simple, three-month endeavor (the "three-hour cruise") that I thought would be a piece of cake, unexpectedly morphed into the massive, expensive, ten-year research and development project that the Global Awakening series has become.

> *I'm really looking forward to hearing what you have to say, because it's really important, and you're the only one who can say it.*

Challenger Peak, after a summer shower.

The Reverend King dream has also helped graphically remind me to hold humility and magnificence—relative and absolute truth, down-to-Earth compassion and lofty wisdom—in both hands, every step of the way, united in my being, in order to do this work with effectiveness and balance. In the dream, being terrified made me feel humble, and being spoken to by one of humanity's greatest lights made me feel magnificent. They are both critically important for us all to hold together in unity as we move forward from here, with our newfound gifts.

I'd like to share the vital, empowering messages of this Reverend King dream as my parting gift here: We each have something important to contribute to the world! We each have a unique gift and our own special voice to share! Now is the time for us all to bring them forth in Joy, to bring them to bear in making a crucial difference at this pivotal moment!

May we all use our newfound power and wisdom wisely; may we all hold the ultimate truth of the fundamental oneness of all things ever closer in our hearts. May we all find our unique gifts and/or voice, and the highest and best ways to contribute them to healing our world. May we all find true joy in all these contributions. May we unite in joyful celebration as we contribute them whole-heartedly in the months and years to come. In so doing, may we all be greatly successful and always able to see the forest thanks to the trees and that big, beautiful blue sky behind whatever clouds may come our way, as we proceed further with effectively healing our world!

And speaking of clouds, whenever I look up into the sweet Southwestern sky and see them now, rest assured: I really do understand that they don't belong to me, at least not in the way I thought they did as a child. Seriously—it's true—after fifty years, I stand humbly corrected! I realize now that my childish, me-first sense of ownership of those clouds was really just a special sort of primitive, foreshadowing premonition of the greater sense of ownership that the trajectory of my life would provide. This is the sort of owning of something that is embodied in the sense of having it be an integral part of oneself—that special shift in consciousness that came about in my life and that I share in many integrative oneness glimpses imparted here and elsewhere in the Global Awakening series.

Maybe there's hope for little Lee's exceptional My Cloud Theory after all.

The End!

CREDITS, ENDNOTES & INDEX

GLIMPSES OF ONENESS

PHOTOGRAPH, DIAGRAM & ARTWORK CREDITS

All photographs, diagrams and artwork illustrations are provided courtesy of the author except: front cover, Shutterstock; 6, H. Emerson Blake; 8-9 underlay, istockphoto; 13 underlay, © 2008 Matthew Hunt/Creative Commons Attribution-Share Alike 2.0 Generic license; 14, © 2013 Max Ronnersjö/Creative Commons Attribution-Share Alike 2.0 Generic license; 15, Michele Rocca, *Angelica and Medora*, circa 1720-1750, Courtesy of Walters Art Museum, Baltimore; 17, Sabra Moore; 18-19 bottom, Lucyna Koch/istockphoto; 20-21, Vic Mansfield, Courtesy of Lee Temple; 25, Jon Sullivan/ Creative Commons CC01.0 Universal Public Domain Dedication; 31 underlay, istockphoto; 39 underlay, © Gary Halvorson, Oregon State Archive; 41, istockphoto; 43, Dennis Neuhaus; 44-45 top, 18th century, Französische Schule, *Trompe l'oeuil*, Courtesy of Dorotheum; 44-45 bottom, © 2010 Arbyfaith/GNU free Documentation License, Version1.2; 46, NASA; 52-53, © 2007 jmblanco 74/Creative Commons Attribution-Share Alike 3.0 Unported license; 54 all, Dennis Neuhaus; 57, © 2013 Henry Mühlpfordt/GNU Free Documentation License, Version 1.2; 59, Emily S. Damstra; 61, © 2007 Michael Graham/Creative Commons Attribution-Share Alike 7.0 Generic license; 63, Emily S. Damstra; 65, © 2011 Øyvind Holmstad/Creative Commons Attribution-Share Alike 2.0 Generic license; 67 top, © 2008 Poliphilo/Creative Commons CC01.0 Universal Public Domain Dedication; 67 bottom, © 2011 David R. Tribble/ Creative Commons Attribution-Share Alike 2.0 Generic license; 76, Lori Nagel; back cover underlay, istockphoto; author photo, Lori Nagel.

ENDNOTES

1. http://www.worldpublicopinion.org/pipa/articles/btenvironmentra/694.php?lb=brglm&pnt=694&nid=&id=)

2. Bob Deans, "This Historic Climate Moment," *Onearth* Magazine, Spring 2013, volume 35, number 1, Natural Resources Defense Council, New York, NY, p. 12.

3. "I met Terry Root, one of the lead authors of the IPCC report, at the Stegner Symposium at the University of Utah (2008). She presented all the IPCC data, and I went up to her afterwards and said, "That graph that you showed, with the possible emission scenarios in the twenty-first century? It looked like the best case was that carbon peaked around 2030 and started coming back down." She said, "Yeah, that's right." And I said, "But didn't the report that you guys just put out say that if we didn't peak by 2015 and then start coming back down that we were pretty much all screwed, and we wouldn't even recognize the planet?" And she said, "Yeah, that's right." And I said: "So, what am I missing? It seems like you guys are saying there's no way we can make it." And she said, "You're not missing anything. There are things we could have done in the '80s, there are some things we could have done in the '90s—but it's probably too late to avoid any of the worst-case scenarios that we're talking about." And she literally put her hand on my shoulder and said, "I'm sorry my generation failed yours." That was shattering to me."

—Tim DeChristopher, Terry Tempest Willams, "What Love Looks Like, A Conversation with Tim DeChristopher," *Orion Magazine*, Great Barrington, MA, January/February 2012, and http://www.orion-magazine.org/index.php/articles/article/6598

4. http://www.refugeesinternational.org/policy/in-depth-report/confronting-climate-displacement?gclid=CNfVvLr17awCFQZeTAodYiPhbA

5. An awakening chronicled in my diary entry in Chapter Twelve.

6. Ken Wilber provides a good explanation of psychological reintegration in his book, *A Brief History of Everything,* on pp. 222–225.

7. For those interested in this approach to self-unification, some great resources to help you walk through the process creatively have recently been published. You can find one source in my reference list: *The Mandala Workbook, A Creative Guide for Self-Exploration, Balance, and Well-being*, by Susanne F. Fincher.

8. Joseph Campbell, *Sukhavati, a Mythic Journey,* (DVD), Joseph Campbell Foundation, 1998, 2005.

SUPPORT MATERIALS

THE GLOBAL AWAKENING SERIES

ACKNOWLEDGEMENTS

GIVING THANKS

Grateful appreciation goes out to all those great beings, places, and things whose direct or indirect help has made me and/or this work and the larger *Global Awakening* project possible: Originating Power, our great Grandfather Fire, the Universal Laws, Space and Time, atoms, the Heavens, the Milky Way, our Solar System and Sun, our Mother Earth, the mountains, especially Arunachala and Challenger Peak, the Thunder Beings, the Cloud People, the San Luis Valley, our combined mountain/valley biosphere, all earthly living things, especially the trees, our garden (food); my loving, supportive life partner; professional associates Bill McKibben (350.org, billmckibben.com), Chip Blake, Madeline Cantwell, and the extended family at the Orion Society (orionmagazine.org/); James O'Dea (jamesodea.com) and the Shift Network (theshiftnetwork.com) and Evolutionary Leaders group (evolutionaryleaders.net); Sidney Piburn, formerly of Snow Lion Publications; Liz Walker (liz-walker.org/), Dennis Neuhaus, Michael Brownlee, Jo Anne Kiser, Paul Shippee, Leigh Mills, Katie Getchell, Grace Anderson, Maria Valdez, Nick Chambers, Mike Wasserman, Tamar Ellentuck, Ceal Smith, Ravie Malhotra, Bob Adler, Mark Jacobi, Pam Gripp, Peter May, Sheila Poor, Bill Sitkin, Sam Cohen, Steve Smolen, Charlene Temple, Nanci Rose-Ritter, Rama C. Hoetzlein, Bill Bauman (billbauman.com), Vamarie Nanej, Buddy Frank (effortlessbeing.com/), Daniel and Vinnie Terres (creativehealthcenter.com/), Mary and Vince Palermo, Kelly and Zana Hart (dreamgreenhomes.com/), Gordon Johnson, Paul Kloppenburg, Everett Buss and Sherry Black, Lynda Kucin, Jack Goldberg, Keith Conway, Talmath Lakai, Richard and Marie-Louise Baker and Christian Dillo of the Crestone Mountain Zen Center (dharmasangha.org/), Matthew Crowley, Kazu Domi and Samuel Sata of Shumei Crestone (shumeicrestone.org/), John Milton (sacredpassage.com), William and Barbara Howell, Kate Steichen, David and Lorain Fox-Davis, Hanne Strong (manitou.org/), Ken Gilbert, Lidian King, Seneca, Plato, Plotinus, Charles Darwin, G.I. Gurdjieff, Alvar Aalto, Colin Rowe, George Hascup, Bernhard Hoesli, Luca Maraini, J. Krishnamurti, Ramana Maharshi, Paul Brunton, Anthony Damiani, Gail Lang, Brother David Steindl-Rast, His Holiness the Dalai Lama, Lakota Chief Arvol Looking Horse, Neil Young, my mother and father and our biological family and all our ancestors.

PROJECT TEAM

The *Global Awakening* series is the result of over 15,000 work-hours spent during the years 2008–2015 by the following devoted group of dedicated individuals: Paul Cash and June Fritchman of Larson Publications (larsonpublications.com), Anne Kilgore of Paperwork, Emily Damstra (emilydamstra.com), Adele Hutchinson (freelancepermissioneditor.com), Cynthia Lane of FirstlightTransformations, Inc. (firstlighttransformations.com), Molly Rowan Leach (http://peacealliance.org/who-we-are/staff/), Sabra Moore, Thayne Rigby (antishrillwebdesign.com/), Kelly Roberts, Dee Rudoff and Elise Rudoff (eliseclaire.me), Lee Temple (primamundi.com), Justin Tribby (jtribby88@gmail.com), and Jelehla Ziemba. Each team member has made significant and important contributions that qualitatively enhance the final work. We hope you will greatly enjoy and benefit from the fruits of our labor!

THE *GLOBAL AWAKENING* SERIES

INFO/LINKS

The *Global Awakening* series integrates widely accepted scientific, spiritual, and environmental perspectives to dramatically unveil our profound collective journey from the Big Bang to the present day—as a necessary primer for understanding, at the deepest core level, who we truly are, what we're made of and how we've come face to face with the greatest human challenge of all time: global climate change. In so doing, it facilitates a large-scale human awareness shift into a new, direct experience of the universe's inherent unity. This enlightening transformation, arguably as essential to successfully addressing the climate crisis as the advancement of green technology or governmental initiatives, inspires broad-based, compassionate, win-win behavior that promotes lasting healing for Earth and her children. For more information, please visit www.primamundi.com

VOLUMES IN THE *GLOBAL AWAKENING* SERIES

(Listings in green are already available or expected soon. All title listings and availability are subject to change without notice.)

SPECIAL FEATURES OF THE SERIES

A PUBLISHING PROJECT THAT STARTED OUT CARBON-NEUTRAL AND ENDED UP CARBON-NEGATIVE

CARBON-NEUTRALITY: You now view on your screen a totally awesome solar-powered clean green document! It was conceived and written entirely on carbon-neutral solar power at the Wingspread Sustainable Homestead in Crestone, Colorado. The Wingspread solar array has also provided ample power to indirectly cover all the necessary and substantial off-site book development and production activities as well. As a result, we have saved more than sixty pounds each of nitrogen oxides and sulfur dioxides, as well more than twenty tons of carbon dioxide from entering the air as a result of all activities directly associated with writing and developing this book (based on EPA electricity footprint calculation models—see www.epa.gov/cleanenergy/energy-and-you/how-clean.html for more information).

GREEN PRESS INITIATIVE: We are committed to maximizing green publishing so as to preserve and protect ancient forests and their vital contribution to Earth's fragile biosphere. In keeping with publishing industry trends that are moving away from physically printed documents, we provide two standard offerings: an electronic e-book, and/or a PDF document.

For those who prefer to have a printed version, our e-book PDF has been formatted so that each e-page is printable 1:1 on a single sheet of 8.5"x11" paper. This allows you to print out as many copies as you desire, with the understanding that you are then completely responsible for the carbon footprint associated with your printing.

This volume also has a custom printing option for this segment of the audience. It is available as a POD (print on demand) book, whose pricing relative to the e-book or PDF versions reflects the additional costs of paper, printing, binding, and distribution to you, as well as the cost of offsetting the carbon footprint associated with all these materials and activities. Please visit www.primamundi.

com/store-freebies/ to see how you can create/purchase a POD copy.

For every 5000 copies of this e-book, we estimate that we save 6 tons of wood use, 49 million BTUs net energy useage, 8,402 pounds of greenhouse gases, over 34,000 gallons of water, over 2800 pounds of solid waste, etc.,   byoffering it electronically as our standard form. These environmental impact estimates were made using the Environmental Paper Network Paper Calculator Version 3.2. For more information, please visit www.papercalculator.org, and www.greenpressinitiative.org.

As a result of these important solar-powered creation and e-book/pdf and Green Press production decisions, we prevent a total of more than two hundred tons of greenhouse gases from entering Earth's atmosphere through the creation, production, and distribution activities associated with this book.

And there's more! To achieve full carbon-neutrality, the remaining emissions associated with the book's printing and distribution (generated by those of us who prefer the POD option) are covered with carbon offsets purchased with a portion of the sales price for printed copies. So as to distribute our offset capital globally, we purchase them from ClimateCare (climatecare.org), and TerraPass (terrapass.com), recognized by offsetconsumer.org and TreeHugger.com, respectively, as two of the world's leading carbon offset providers.

A CARBON-NEGATIVE PROJECT!: Finally, to become carbon-negative (meaning that we actually contribute to a healthier atmosphere through the creation of this book), the author and the book's development company, Shining Golden Suns, LLC, have, prior to its release, caused three hundred trees to be planted in North America, and over an acre of virgin Central American rainforest to be permanently protected on behalf of this enterprise. To our knowledge, this is one of, if not the world's first carbon-negative book project. Please visit www.arborday.org to learn how you too can contribute to a healthier biosphere.

LEGAL DISCLAIMER/CAVEATS

I've written and assembled all the Global Awakening documents in general and this book in specific for our general entertainment, education, and amusement. Its creation has been a long and mostly personal inspiration that has also been combined with the important contributions of people mentioned herein. I've received no personal compensation from any source for the preparation of this material. To the contrary, I have contributed large amounts of my own time, energy, and capital toward its creation, as my own personal sort of "gift to the world" in these challenging times. Any remuneration and/or royalties that I may receive will go first to repaying my own substantial contribution, and then toward worthy enterprises of my choosing.

The views expressed herein are mine and mine alone (unless quoted or noted otherwise), and do not represent the views of any organization or person with whom I am or have ever been, or not been, affiliated.

As an independent writer/producer of this material, I am in no position to make any warranties of appropriateness or fitness to anyone, for any purpose. You have decided to purchase and read this material (thank you kindly!) of your own free will, and you understand and agree that I am not forcing you to do so, or to do anything that you do not wish to do of your own intelligence and/or volition.

Producing a project such as this involves, as you might imagine, immensely time-consuming, intricate, and detailed work over extended periods of time. Information has been culled from a tremendous variety of sources (see the endnote list, and/or the References and Resources volume of the complete series, for a sense of just how much has been involved). Every effort has been made to provide the most accurate information available at the time, and to correctly credit and include all sources of information and their authors who have found their way into these final documents.

Still there may have been unintentional errors, inaccuracies, or omissions in assembling and presenting this material to you. If, in reading it, you come across such an error, inaccuracy, or omission—for example, a reference, idea, etc., that is inaccurate, incorrectly credited, or not been credited at all, and so on—please notify the publisher (Shining Golden Suns, LLC, P.O. Box 220, Crestone, Colorado 81131) so that it might be swiftly corrected.

Finally, wholly new and/or updated versions of existing issues may be released from time to time, as they are finalized and made available. We bear no responsibility for notifying any customers of these new issue releases. Please visit www.primamundi.com to learn about all related current events and updates.

ABOUT THE AUTHOR

Lee Temple is an author, visionary, elder, community organizer, and global sustainability activist/consultant who lives the low (carbon) life in the remote, high-mountain desert of south-central Colorado. Together with his life partner, Lee designed and built the Wingspread Sustainable Homestead—a comprehensive demonstration of how a typical American family can affordably reduce its overall carbon footprint by more than 75%—by relocalizing their energy, food, and livelihood—and adopt an exciting, enjoyable, fulfilling, and enriching way of life in the process. He has generated his own solar power for more than two decades.

Lee is a retired architect/university professor (he taught architecture and urban design courses at Syracuse, Hobart/William Smith, and Cornell Universities, and headed his own architectural firm for over twenty years).

He has made spiritual pilgrimages to Europe, India, the American Southwest, Easter Island, and Hawaii. He has received many spiritual teachings, transmissions, empower-ments, and initiations, including the Kalachachra initiation that was given in 1991 by His Holiness the Dalai Lama in Madison Square Garden, New York.

Lee co-founded the Crestone Sustainability Initiative, whose goal has been to create a comprehensive sustainability policy for reducing Crestone's carbon footprint significantly by 2020, and to make it a working model for other towns in the southwest. In addition to his engagement in various creative, spiritual, educational, sustainability, and philanthropic activities over the past thirty years, Lee manages Shining Golden Suns, LLC, the developer of the Global Awakening project. Learn more at www.primamundi.com.

ENDORSEMENTS

Lee Temple is a modern day Renaissance man. His command of so many fields such as cosmology, science, philosophy, psychology, spirituality, ecology and systems thinking is truly impressive. He puts all this knowledge together in a lyrical, informative and practical presentation designed to provide us with all we need to make informed choices and take wise action to save our planet.

His mapping has a moral imperative: we must transform our consciousness and act.

JAMES O'DEA
AUTHOR OF *Cultivating Peace: Becoming a 21ˢᵗ Century Peace Ambassador*

Over the past two decades, I've personally witnessed Lee Temple become a powerful spiritual visionary, an accomplished mystic and a highly-regarded and well-respected sustainability elder. Today he is an important spokesperson to humanity for the heart and soul of the Earth, a valuable guide for the Earth-healing movement that is happening in all parts of our world.

Lee has generously chosen to share his wisdom and unconditional love through this inspiring and comprehensive "Global Awakening" Series—a compelling work that draws us pragmatically, intellectually, energetically, and spiritually into the truth of our unity and wholeness. He uses the language and lenses of science to reveal the essential nature of the universal creation process, so that we might better understand the magical, miraculous spirit embodied in its mechanics and behavior—the spirit that is the bedrock foundation of humanity and our world.

Lee's full body of work and this powerful volume artfully and authentically capture the soul of existence and the spirit of aliveness that all things share. It's a personal invitation for us to remember the multi-dimensional nature of all creation and our ever-evolving participation in it, so that we might joyfully join with others to enter consciously into the magic of creating a whole new human presence on Earth—one centered in authentic global Earth-healing and stewardship, one that is, at its core, profoundly respectful of the deeper unity of all life and our Earth mother herself.

Lee shows us, through his own in-depth experiences and many other compelling examples from all around the world, that these emerging ways of honoring all of creation can deeply ground and connect us to each other and our common, universal source of existence.

As Lee beautifully explains, from this level of our oneness with others and all that is, we will find deep and comprehensive answers to the current environmental crisis of global climate change—answers that will help align us more effectively with the eternal life and flow of the universe, as we consciously and compassionately join together to heal our world.

BILL BAUMAN, PH.D., FOUNDER, CENTER FOR SOULFUL LIVING, INC. & WORLD PEACE INSTITUTE, INC.

Lightning Source UK Ltd.
Milton Keynes UK
UKHW052256230920
370410UK00012B/168